Springer Polar Sciences

This Series is now being indexed by SCOPUS. *Springer Polar Sciences* is an interdisciplinary book series that is dedicated to research in the Arctic, sub-Arctic regions, and the Antarctic. In recent years, the polar regions have received increased scientific and public interest. Both the Arctic and Antarctic have been recognized as key regions in the regulation of the global climate, and polar ecosystems have been identified to be particularly susceptible to the ongoing environmental changes. These changes are having widespread implications for human communities, businesses, and governance systems and are interacting with demographic shifts, globalisation, resource development, cultural change, territorial disputes, and growing calls for self-determination in some regions. Consequently, the international efforts in polar research have been enhanced considerably, and a wealth of new findings is being produced at a growing rate by the international community of polar researchers and those who live in the region.

Springer Polar Sciences aims to present a broad platform that will include state-of-the-art research, bringing together both science, humanities, and perspectives rooted in indigenous and local knowledge to facilitate an exchange of knowledge between the various polar science communities. The series offers an outlet to publish contributions, monographs, edited works, conference proceedings, etc. Topics and perspectives will be broad and will include, but not be limited to, climate change impacts, climate change policy, environmental change, polar ecology, governance, health, economics, indigenous populations, tourism, resource extraction activities, and research design in polar regions. Books published in the series will appeal to scientists, students, polar researchers, community leaders, and policy makers.

P. J. Capelotti

Lost Greenland

The American desire for an Arctic colony, from Benjamin Franklin to Donald Trump

Springer

P. J. Capelotti
Professor emeritus of Anthropology
Penn State University
Philadelphia, PA, USA

ISSN 2510-0475 ISSN 2510-0483 (electronic)
Springer Polar Sciences
ISBN 978-3-032-27840-1 ISBN 978-3-032-27841-8 (eBook)
https://doi.org/10.1007/978-3-032-27841-8

This Springer imprint is published by the registered company Springer Nature Switzerland AG
The registered company address is: Gewerbestrasse 11, 6330 Cham, Switzerland

To C.L., with love

Preface

The goal of this work is to provide an accessible historical framework in which to understand the current American obsession with Greenland. It is an extension of historical archaeological research I have conducted in the circumpolar world for nearly half a century. My focus during that time has been on the archaeology of polar explorers, particularly in Svalbard and Franz Josef in the European Arctic and Labrador in the western North Atlantic, as well as military operations in and around Greenland during the Second World War. So when I turned to the larger scope of questions of Greenland and its historical relations with the United States, and Greenland in relation to Scandinavia, and Greenland in the emerging global resource competitions, there were many additional layers one was required to cover, including Greenlandic prehistory, the history of science, diplomatic history, economic history, and the American treatment of indigenous populations. No single work can adequately address each of these topics, which in themselves have produced libraries full of studies over many years. This work, therefore, is an attempt to crystallize the history of the American interest in the largest island in the world, beginning with the person of Benjamin Franklin, who with Thomas Jefferson set so much of the stage for America to walk out upon.

Philadelphia, PA, USA
February 2026

P. J. Capelotti

Acknowledgements

For permission to use imagery, the author wishes to thank Ms. Andrea Krausmann for the photograph of her father during the Second World War; the Midway Village Museum for images related to the flight of the *City of Rockford* and its location and removal from Greenland 40 years later; the Ward M. Canaday Center for Special Collections of the William S. Carlson Library at the University of Toledo for use of pages from William Carlson's Greenland notebooks; and the American Geographical Society Library at the University of Wisconsin-Milwaukee Libraries for permission to use maps from Louise Boyd's expeditions to East Greenland. Other imagery not from the author himself is gathered from public-domain sources and cited as appropriate in the individual credits. If any of these are incorrect, please contact the author via the publisher. The author would like to thank his many colleagues who have provided advice and support over many years. For a July 2025 visit to Bluie East 2, particular thanks are due from the author to Mateo Caramés, Niccolo Segreto, and Anganni, all of Nunatak Adventures, as well as Jakob Ibsen of Hotel Kulusuk.

Competing Interests The author has no competing interests to declare that are relevant to the content of this manuscript.

Introduction: Gamatron

In the late summer of 1941, a young automotive worker named Thaddeus Novak was laid off from the assembly line of the Chrysler Motors Dodge main plant in Detroit. On Monday, August 25, 1941, he stopped at the factory to say good-bye to friends and to return his employee identification badge. Later that day he was sworn into the U.S. Coast Guard. Three days on a train took him to the Coast Guard's training station in Algiers, Louisiana, across the Mississippi from New Orleans. After basic training, Novak served at lifeboat stations on the Great Lakes and then the Coast Guard Yard at Curtis Bay, Maryland. In the summer of 1942, he was transferred to the Coast Guard Receiving Station in Boston, Massachusetts. There, Novak joined the great American rush to defend the island of Greenland that followed the German invasions of Denmark and Norway in 1940.

In Boston, Novak boarded the *Nanok*, one of several small vessels that looked like "huge wooden shoes, and were needed because they were capable of snaking-through narrow Greenland fjords, dense fields of icebergs and cake ice. Their maneuverability could not be equaled by larger, freighter-sized vessels. For the most part, large freighters were to bring supplies from the U.S. to Greenland, but the wooden shoe trawler fleet would do most of the distribution to the many small installations. Despite being relatively petite in size the trawlers were capable of ferrying some 90 tons of hold cargo and many more tons on deck. It seemed most probable that enemy submarines would so treasure their secret location that they would even refrain from surfacing to deck-gun a tiny trawler into oblivion, which of course they could. The wooden shoe trawler's true enemies were the elements. In all but one instance, they would prove to be more than worthy adversaries."

Upon arrival in Greenland from a long cruise from Boston, *Nanok* was escorted up the immense length of Tunulliarfik Fjord by the U.S. Coast Guard cutter *North Star*. A ten-year-old Arctic supply trawler originally on service with the Bureau of Indian Affairs, *North Star* served just prior to hostilities as a support vessel for Richard E. Byrd's third Antarctic expedition. One hundred feet longer than *Nanok*, *North Star* plowed ahead towards the epicenter of American military power in the North Atlantic at the start of the war, Bluie West 1, near the native Greenland village of Narsarsuaq.

Bluie, the military code name for Greenland, functioned first as a finding aid, with the Bluie bases identified by location and number on either the east or west coast of Greenland, with Cape Farewell at the very southern tip of the island as fulcrum. It was also a simple alternative to American G.I.'s inability to pronounce Greenlandic place names. If "Bluie" proved too much, one could simply say the letter "B", while other bases added still more recognizable recall code names: in the case of Bluie East 2, for example, it was "Optimist."

Novak was instantly awed by the sheer immensity of the place.

> The fjord's jagged shoreline gradually magnifies into giant mountains on both sides of the fjord. As we proceed, the mountains rise higher and higher to breathless heights and from their crests hang giant icicles with peels of ice-cake frosting. The fjord is a narrow, water-filled, bottomless valley between snow and ice-crowned peaks of multi-colored rock. The mountainsides plunge downward at steep angles. Even small beaches are rare, and most have Eskimo villages atop them. Probably because fish and birds and seals abound in the vicinity in abundance enough to sustain life.
>
> There is a majestic splendor to the mountains as we glide between them. Their magnitude fills me with awe, and their silence enhances a sense of loneliness. The *Nanok* is a mouse creeping into a crevasse. As we progress slowly up-fjord, we learn we are bound for Narsarssuak, a command center for the operation of naval patrol planes. There are at least 400 men including Army, Navy, Coast Guard, and civilian construction personnel. It is also the location of a large army base center. The installation's code name is BW 1 (Bluie West 1) (Capelotti, ed. 2005: 47).

Evincing every bit of the fable of the adaptability of the American enlisted man, Nowakowski learns more from his first brief encounter with a Greenlander than all the Norse, Danish, British, and American officers and explorers had learned in a thousand years of exploration, exploitation, and colonization: "No longer must I wonder what I may teach these uneducated natives, but what they will teach me…"

> His conveyance is a kayak some fourteen to sixteen feet in length, about two feet in width and perhaps eighteen inches in depth. It tapers fore and aft to pointed bow and stern. Its fuselage is a fragile-looking wooden skeleton covered with tautly-drawn seal skin. The cockpit is circular, precisely amidships, and is barely large enough to accommodate one person. It has a flexible, collapsible collar of the same type skin. A most uncomfortable appearing craft. Its pilot must wriggle into it as perhaps a woman might wriggle into a girdle. He must sit flat with his legs stretched straight forward, unbent. In a very short time the legs must become numb. Once inside the cockpit, its skin collar is raised and drawn tight around the passenger's waist with drawstrings, making the vessel watertight. Also, the passenger wears a waterproofed sealskin jumper that tucks into the cockpits collar before it is tightened around the waist. The jumper has an attached hood that is drawn snugly around the neck with drawstrings. Another cord vertically encircles the entire face. Yet another cord horizontally encircles the forehead level of the head. The passenger thus becomes totally waterproofed within a sealskin cocoon.
>
> My visitor laid aside his long, slender paddle that had a flat blade at each end and untied his cockpit collar. From somewhere inside his cramped quarters he withdrew a small codfish and held it up for my approval. I smiled and nodded approvingly. "Nice fish." I said, doubtful that he understood. He made a motion toward his mouth and away, and back to his mouth and away again. For a moment I thought he was trying to let me know the fish is good to eat. Then I recognize an old and common American habit. He slaps at his heart as if he is searching for something. Of course! He is trying to let me know that his inner blouse pocket

> is empty of cigarettes. Motioning the fish back and forth, to and away from his mouth, inhaling and exhaling, was his way of demonstrating the act of cigarette smoking! Damned clever of him.
>
> To verify whether I was correct in my assumption, I pulled out a pack of smokes. I shook a cigarette part-way out of the pack and extended the pack toward the visitor. His golden-tan, moon face glowed when he saw the butts. Instead of taking one as is our custom, he grasped the entire pack and pulled it out of my hand and replaced it with the codfish!
>
> "Hey!" I began, but away he went, bound for the village across the fjord. He paddled swiftly without a backward glance.
>
> Standing in line with the rippled wake of the small vessel, I wondered if this was Greenlandic custom. There was no doubt I had been "taken" by this iceberg carpetbagger. He knew I was unable to pursue him. My spontaneous anger turns into laughter (*ibid.*: 49–50).

Novak's diary is one of the few first-hand accounts to survive from the Greenland Patrol, and the only such first-hand account from the perspective of an involved young, enlisted seaman (Fig. 1). The reason for this lack of documentation is clear enough. Keeping of any such journal in wartime was strictly forbidden, and Novak came within an ace of being cashiered on account of it. Its survival therefore occupies a unique place not only in the history of the U.S. Coast Guard and of U.S. naval operations in the Arctic and during the Second World War but stands as a remarkable document in the nearly 300-year American fascination with the largest island

Fig. 1 Thaddeus Nowakowski (Ted Novak) as a U.S. Coast 3rd class petty officer following his return from Greenland in 1943 (photo courtesy of Andrea Krausmann)

in the world. It offers a striking account of the wartime encounters between native Greenlanders and enlisted sailors and their officers, and between various branches of American services in the far north. It also stands in stark contrast to the inveterate ethnocentrism and racism of established American scholars and explorers from Samuel Morton to Robert Peary.

By early September, *Nanok* was anchored about 60 miles south of Bluie East 2, as Novak and his mates hauled two heavy motor sledges up towards the Greenland Ice Cap. "No one on board or ashore has ever seen one before. Something new for the Arctic and Antarctic. After getting them ashore, we must carry them up a steep hill of snow and rock. My legs sink in snow halfway to the knees and twist and slide on small sections of ice. It is a travel of several inches at a time. An army lieutenant named Max H. Demorest follows our movements as though the sleds were priceless. I ask his interest and learn that he travels the Ice Cap often on skis or by dog sled. To him the motor sleds are a Godsend. He says there have never been any motorized sleds in Greenland before. As matter of fact, these are the only ones he has ever heard of" (Capelotti ed., 2005: 49–50).

Max Demorest was a Princeton Ph.D. glaciologist from Flint, Michigan, who had developed a theory of glacial ice crystallization and re-crystallization that became widely accepted as the basis of glacial flow. Little more than two months later, when a plane flying from Iceland to Greenland crashed on the Ice Cap, a second plane, a Boeing B-17 Flying Fortress bomber sent out with other searchers, also crashed in the Comanche Bay area. For nine days the injured crew were imprisoned in the cabin of the bomber. Demorest and his new motor sledge would attempt to rescue them.

Soon after his meeting with Demorest, Novak found himself alone on *Nanok's* deck, standing the midwatch, in despair at his loneliness away from home. It is here again that Greenland comes to his rescue.

> I am alone in onyx darkness. Aurora borealis, sensing my mood, presents a special performance for me. Mountainous globules of multi-colored paint pours down from the heavens in a straight line from horizon to horizon. Colors are exceptionally brilliant this night. They drip and sag as liquid pouring down an invisible pane of glass across the sky above me. The colors appear to be reachable by hand. Icebergs and sea reflect the colors above. It is a wondrous sight! Patterns and tones are ever-changing. Colors fade and are replaced by other phenomena I call "slow lightning." In the black sky, "slow lightning" does indeed resemble lightning in slow motion. It constantly changes direction at varied slow speeds. It hovers many seconds at a time, suspended in still, chill air. I have never heard it described before. After enjoying the display for quite some time, I feel reconditioned… (*ibid.*, p. 99).

On the 17th of September, *Nanok* delivered supplies and towed a small barge to Bluie East 2, but it is unclear if Novak himself went ashore during the brief visit. Had he done so, it is fanciful although hardly inconceivable that he would have been in a position to drive a Dodge truck which, just a year earlier, he had helped to assemble. By the time *Nanok* had reached the East Greenland coast, the Bluie East 2/Optimist base was already home to a sizable fleet of American vehicles. These vehicles and the generators, aircraft, ships, and tens of thousands of fuel drums that fueled them were by then being produced at lightning speed by the same assembly lines in Michigan and throughout America that had made workers like Novak surplus to requirements just a year earlier (Fig. 2).

No less than 14 bases were established by the U.S. military in Greenland in the early stages of the Second World War, at least 17 by the end of the war. Many of these were constructed by private contractors operating under the direction of the U.S. Army. U.S. Navy Naval Construction Battalions (Seabees) did not exist prior to their formation in early March of 1942. Greenland naturally figured as one of the first of their many wartime challenges. Wharves were created virtually overnight at some of the most remote and treacherous landing spots on the planet by the simple expedient of building a cradle of crossed beams and then filling the empty cradle with nearby rocks and stone. At least these so-called "crib piers" were simple if *you* were not one of the sailors shouldering the heavy beams from ship to shore, as no such heavy timber exists anywhere in Greenland.

When one surveys an abandoned Greenland base such as Bluie East 2, about 25 nautical miles northeast from the airport at Kulusuk, the first impression is of the hundreds of fuel drums (that recently numbered in the tens of thousands until a much-belated clean-up began in 2018), followed by wonder at a vast landscape of abandoned heavy trucks, light trucks, trailers, steamrollers, cranes, even, under a

Fig. 2 Foreground: remains of an apparent 1942–47-ish Dodge 1/2-ton flathead. Under the brands of both Dodge and Fargo, more than 335,000 of these light trucks were produced during WW2 and this example, at the site of the former Bluie East 2 base, is an archaeological reminder of the American rush to defend Greenland after the German invasions of Denmark and Norway in 1940. In the background are the remains of apparent U.S. Army 2-1/2-ton dump trucks (photo by the author, July 18, 2025)

welter of debris, what is likely one of the earliest surviving radiosonde tracking stations.

At Bluie East 2, no less than two dozen heavy trucks are in evidence, and this reflects just some of the basic kit of a naval construction battalion as it became formalized by the end of the war. In trucks alone, such an advance base required no fewer than 77 vehicles, including "a 3/4-ton ambulance, eighteen 1/4-ton reconnaissance cars (jeeps), and four 3/4-ton weapon carriers. The heavy trucks included one 2^{1}/2 -ton, six-by-six, 750-gallon tank truck; four 1^{1}/2-ton, six-by-six, cargo trucks; four 2^{1}/2-ton, six-by-six, cargo trucks; thirty-two 2^{1}/2-ton, six-by-six, dump trucks; one 2^{1}/2-ton, oilfield body truck, and one 4-ton, six-by-six, cargo truck. In addition to the trucks, there were 20 trailers, of various types, and one fifth-wheel, 2^{1}/2-ton, six-by-six truck tractor" (Navy Department, Washington, 1947: 152).

William S. Knudsen, recruited from the automobile industry in 1940 by Franklin Roosevelt to accelerate production of war vehicles and materiel to levels never before achieved, later remarked, "We won because we smothered the enemy in an avalanche of production, the like of which he had never seen, nor dreamed possible" (Reed, 2022). Knudsen's role in ramping up American production was so critical to the defense of Europe that he was featured on the cover of *Time* magazine in late 1940 and decorated by the Kingdom of Denmark in the war's immediate aftermath. These vast amounts of industrial production left an archaeological record across the globe and nowhere so prominently as in Greenland.

By the last week of November, as it again rounded the southern tip of Greenland, this time westwards, the *Nanok* hurriedly delivered supplies to other remote stations prior to an anticipated rush back to Boston to avoid being trapped for the winter.

Twelve months earlier, the U.S. Coast Guard cutter *Algonquin* had delivered a radio beacon ashore at Frederiksdal (now Narsaq Kujalleq) near Cape Thorvaldsen in southernmost Greenland. Construction of this long-range navigation station continued throughout the winter, with the cutter *North Star* landing U.S. Army staff as the station became operational in January of 1942.

Now, in late November 1942, as the ice closed around them, Novak and his mates on *Nanok* loaded cargo for delivery to this small outpost, "a place called Gamatron, wherever or whatever that is" (*ibid.*, 151). Novak was not to know it, but the islet upon which the ultra-modern Gamatron had been sited lay less than two miles from where Norse colonist Herjólfr Bárðarson had settled his clan as part of Eiríkr the Red's Greenland colonizing expedition of 985.

> The *Nanok* is backed into a short, dead-end fjord off a larger, main fjord. The fjord is surrounded on three sides by mountains that rise straight up. The fourth side has a small, narrow beach onto which we offload. An anchor has been dropped forward of the ship and a line from the stern has been taken ashore some seventy feet away. This seems to be quite a common tie-up arrangement in short, dead-end fjords.
>
> It looked like a big blow coming up so Maggie wanted a second stern line tied ashore. To do so, we must first shift location of the first line. It had been fouled around large rocks by vicious but small waves. There is no end of trouble disengaging it. In the end an ax had to be used to chop some of it away. When settled, another scow load is off loaded ashore and we knock off for the day (*ibid.*: 152–153).

Another American who experienced Greenland firsthand during the war and left behind fragments of testimony was just about the polar opposite of Ted Novak. Pierre Boucheron landed in Greenland in the late fall of 1941 and remained throughout that winter, not as a young enlisted sailor but as a 53-year-old U.S. Navy Reserve officer with experience in marketing, selling, and servicing radio technology.

Like Novak, Boucheron kept a log of his year in Greenland and, like Novak, tried and failed after the war to work his experiences into a completed manuscript acceptable to a publisher. As other writers like Norman Mailer and James Jones experienced international post-war acclaim writing of their experiences in the Pacific, it would take 35 years before Sloan Wilson, famed author of *The Man in the Grey Flannel Suit*, published *Ice Brothers*, his fictional account of his own service on the Greenland Patrol. It was not until seven decades after the end of the war that Boucheron's grandson uncovered his Greenland notes and published excerpts online (Boucheron, 2015).

It was Pierre Boucheron who led the 14 men who set up the station that became known as Gamatron. It is also probable that Boucheron himself gave the station its unusual name. As his grandson percipiently noted: "The name 'Gamatron' sounds suspicious. It does not resemble Danish or Inuit names in Greenland. It looks like a combination of 'gamma ray' and 'electron.' 'Gammatron' was the trade name for a vacuum tube made by Heintz & Kaufman in San Francisco, used in transmitters for the Dollar Steamship Company's fleet in the Pacific. Pierre probably knew about the vacuum tube. Did he rebrand the island? If so, the name stuck, as it appears in other accounts of the period and on maps" (*ibid.*).

In Pierre Boucheron's telling, he had been offered two active-duty posts prior to the U.S. entry into the war but turned them down, feeling himself unqualified. When offered a "last chance" assignment to establish Gamatron, he took it.

> I look in the files of the Navy Department, the U. S. Coast Guard (normally the watchdog of the North Atlantic on its ice patrols), the National Geographic Society's file of published data on Greenland, and finally Admiral Byrd and his key officers on prewar Antarctic expeditions. I inherit some of that gear and later find it useless. I meet Marconi civilian engineer Griess of Chelmsford, England, who will accompany me to set up the Marconi Adcock Strategic Direction Finder.
>
> The Greenland assignment, July 1941 to October 1942 … was my first independent naval command… Our band is the usual conglomeration of personalities, temperaments and origins that one finds in any American group. Beginning with the Commanding Officer who is French-born, we have two of Italian descent, one Dutch, four Irish, one Pole, one Czech, two Germans, several Scandinavians, and a British subject, Mr. Griess. A word about him.
>
> Mr. Griess is the Marconi engineer. Slight of build, he is aesthetic and cultured, with the most perfect Oxford accent. He is at once a delight and a revelation to us. By his constant good cheer and untiring energy, he has shown the way to many a job that looked impossible. His example has made a few of the boys snap out of it, boys not accustomed to hard outdoor work, boys who might not otherwise realize that we are at war, that every man must do his share for the common good (*ibid.*).

Thus did Gamatron, on a small islet in southern Greenland, with its futuristic science fiction name and connections to Old Norse history, become perhaps the first such multi-national agglomeration to not only resemble but predict what after the war would eventually become the North Atlantic Treaty Organization. But all that lay in the future. In early December 1942, panic began to set in among Novak's mates on the *Nanok*, who saw dwindling chances of getting home before the freeze.

"Surely, if we are to be stateside by Christmas, we should be on our way by now. I am deep in gloom but not alone. Many others mention similar feelings to one another" (Capelotti, *op cit*: 153). On December 4th, the crew learned further depressing news: Lieutenant Demorest had fallen into a crevasse on the Ice Cap while on one of the motor sleds they had delivered to him. "He was a very fine gentleman we will all miss very much" (*ibid.*: 156).

On November 28, the Coast Guard cutter *Northland* received word that the U.S. Army had made contact with the crew of the downed B-17 Flying Fortress on the ice. Using his new motor sledge hauled ashore by Novak and his mates, Lieutenant Demorest arrived in the vicinity of the crashes and then reached the missing aircraft on foot. When he attempted to bring up the sledge, a thin snow bridge could not bear the sledge's weight and Demorest fell to his death at the bottom of the deep crevasse, aged 32.

From *Northland*, 28-year-old Lieutenant John A. Pritchard and his 29-year-old radioman, Benjamin A. Bottoms, took off in a Grumman J2F-5 Duck amphibian pontoon aircraft and made a wheels-up landing on the ice cap, some four miles from the wreck of the B-17. Pritchard managed to shepherd the three injured survivors back to the Duck, as Bottoms maintained radio contact with *Northland*. The weight and take-off limitations of the Duck forced Pritchard to leave one of the airmen behind until the next day. After he and Bottoms returned and successfully retrieved the man, worsening weather apparently forced Pritchard off course. Bottoms' radio signals to *Northland* grew faint and then went silent.

For a month, volunteers from *Northland* searched for Pritchard's lost flight without success. An army aircraft located the wreckage four months later, but no bodies were ever found. Pritchard and Bottoms were posthumously awarded the Distinguished Flying Cross. A memorial plaque to the three lost servicemen is fixed to a large boulder near Kulusuk harbor. Demorest is remembered on a monument to the missing at the Cambridge American Cemetery in Cambridgeshire, England.

On December 5th, to their fury, the *Nanok's* crew was again assigned delivery patrol. "The *Nanok's* booms were rigged out in preparation to onload a shithouse-full of barrels of fuel for Gamatron again! This was not supposed to be our duty but someone higher up changed many plans" (*ibid.*). By the first anniversary of the attack on Pearl Harbor, Hawai'i, *Nanok* was back at Gamatron with its load of fuel drums. "Gamatron's base commander estimates it will take at least a week to offload our cargo. Maggie says if it takes that long we will miss whatever opportunity remaining to return stateside before a solid freeze-in" (*ibid.*: 157).

Novak suggested to his skipper, an Icelandic salt hurriedly commissioned into the Coast Guard Reserve by the name of Magnus G. Magnusson, that the crew save time by rigging a kind of clothesline between their ship and the shore at Gamatron and simply float the lighter-than-seawater barrels of diesel ashore one at a time, rather than trying to load them onto small barges and then having to navigate the unwieldy platforms across the rough surf to the beach. Magnusson eventually agreed, and a job expected to last at least a week the *Nanok* crew finished in less than a day.

One week later, back at Bluie West 1, *Nanok* was joined by sister ship *Natsek* and their escort to Boston, USS *Bluebird*. Blinded on the return voyage by ice in the Strait of Belle Isle off Newfoundland, *Natsek* went down with all hands during the night of 16–17 December 1942 and was never seen again. Unlike *Nanok*, commanded by Magnusson with his decades of North Atlantic experience, *Natsek* had been under the command of Lieutenant (j.g.) Thomas Sargent La Farge. According to the official U.S. Coast Guard history of the Second World War—and in an unselfconsciously ironic phrase typical of the Second World War—La Farge had joined the Coast Guard "to paint, while performing regular war duties" (U.S. Coast Guard Historical Division, 1945: 44).

By the end of 1942, both Boucheron and Novak had returned safely to the United States. The impression Greenland left upon these Americans, profound and lasting as it was, would be dwarfed by the sheer tonnage of material culture left upon the Greenland landscape by the participants in the Second World War and a host of other Americans who experienced the great island both before and after them. As a result, there exists in Greenland a dramatic archaeological and anthropological database.

Of the latter, there were the intimate relationships between Americans and Greenlanders from at least Robert Peary and Matthew Henson onwards that were frowned upon, if not forbidden, by Danish authorities, but inevitably impossible to prevent. Greenland was ground zero in the fall of 1909 for the Cook-Peary feud over American priority at the North Pole, a battle royale that reshaped American media for all time. Peary and his African-American compatriot Matthew Henson both fathered children with Greenlandic women, and these children, as octogenarians, were brought to Washington, D.C. in 1987 in order to participate in Henson's reinterment next to Peary's grave at Arlington National Cemetery in 1988.

In the summer of 1933, Charles and Anne Morrow Lindbergh crisscrossed the Ice Cap during their 30,000-mile survey for potential Pan American Airways Trans-Atlantic air routes. The Greenlanders named the Lindbergh's Lockheed Sirius floatplane *Tingmissartoq, 'the one who flies like a bird'*, and on August 20, 1933, Anne Lindbergh asked a Greenlandic boy, Levi Imaka, to paint the name on the aircraft nose. The aircraft is preserved today in the Pioneers of Flight gallery at the National Air & Space Museum just a few feet from Lindbergh's *Spirit of St. Louis*. Peary, Cook, Henson, the Lindberghs, famous as they are, barely touch upon the complex historical and cultural connections between the United States and Greenland and between Americans and Greenlanders.

Contributions to this record of the American presence in Greenland derived as well from numerous other explorers and service personnel, from Elisha Kent Kane, Isaac Israel Hayes, and Charles Francis Hall, to Evelyn Briggs Baldwin and Parker "Shorty" Cramer and Bert "Fish" Hassel, to scientists such as the University of Michigan's William Herbert Hobbs, to the post-war and Cold War bases, both successful (Thule/Pituffik) and disastrous (Camp Century). These Americans were assisted at virtually every step by both native Greenlanders and by Danes, Icelanders, and Norwegians who actually understood the Arctic.

In addition to Sloan Wilson's *Ice Brothers*, the island has figured in American popular culture from the aviation novels of Ernest K. Gann to an early twenty-first-century John Griesemer tale about a man mistakenly assigned to the far north called *No One Thinks of Greenland*, a title that would not have survived the election of Donald Trump. In Gann's *Island in the Sky*, Air Transport Command (ATC) pilot Dooley is captain of a four-engine aircraft flying cargo between Greenland and North America, when icy conditions force the aircraft down in sub-Arctic Canada. When the 1944 novel was made into a motion picture a decade later, John Wayne was cast as the experienced Arctic flyer Dooley. In Ben Stiller's 2013 remake of *The Secret Life of Walter Mitty*, the exhausted corporate drone Walter makes the first truly courageous act of his life by leaping aboard a nearly airborne helicopter in Greenland.

As for the archaeology, it consists of nineteenth-century shipwrecks and the northernmost marked grave of an American (along with the unmarked graves of explorers and servicemen like Max Demorest lost into the deep recesses of the ice). One finds depots of coal to fuel the ships of polar explorers and the raw materials cached so that an early American explorer could generate hydrogen for his Arctic balloons. There are the abandoned sites of the then-ultra-modern wartime radio direction finders like that at Gamatron, and the sites of the post-war Distant Early Warning (DEW) Line system. There is a long-abandoned under the ice nuclear missile base slowly flowing with the movement of the ice into the sea, where it is expected to arrive, radioactivity and all, later this century. There are the tens of thousands of fuel drums delivered to remote sites like Gamatron and Bluie East 2 by young sailors like Ted Novak, sources of localized contamination that went ignored by the Americans for decades until the Danes finally stepped in, albeit reluctantly, to clean them up. There is the stupendous array of vehicles and facilities orchestrated during the war by William Knudsen and the entire force of American production. These, in turn, in the aftermath of war, would slowly transform the cultural infrastructure of the great island into a somewhat loosely connected transport system.

Greenland, since the 1720s under the sovereignty of a Dano-Norwegian kingdom that claimed land rights tracing to the voyages of Eirík the Red, in 1979 became an autonomous territory of the Kingdom of Denmark. Yet it has also seen Americans come and go since none other than Ben Franklin turned his attention to the place in the 1730s, just a decade after its absorption into the Danish crown. It was coveted by the Secretary of State who purchased Alaska. It was ignored by a president who wanted from Denmark not icy mountains but a submarine port in the Caribbean. It was eyed for purchase by another American president emboldened by his vast powers in the aftermath of the Second World War. In 2025, Greenland again emerged from its relative northern obscurity to dominate global news in ways unthinkable since Cook and Peary in 1909, by an American president every bit as skilled in media manipulation as was Robert Peary.

Now, in a warming world, Kalaallit Nunaat, the Land of the Greenlandic Inuit, has permanently emerged from a millennium in the shadows of world history.

References

Boucheron, R. (2015). *Greenland assignment*. Published on-line at: https://acrossthemargin.com/greenland-assignment/

Capelotti, P. J. (Ed.) (2005). *Life and death on the Greenland Patrol, 1942*. Gainesville: University Press of Florida.

Reed, L. W. (2002). The amazing story of America's War Production. *Times-Herald*, online at: Retrieved February 28, 2026, from https://www.times-herald.com/opinion/the-amazing-story-of-america-s-war-production/article_ed5ea76f-230b-597b-841a-cbec81ec9e37.html.

U.S. Coast Guard Historical Division. (1945). *The coast guard at war, Vol. II: Greenland Patrol*, Washington, D.C.: Government Printing Office.

U.S. Navy Department. (1947). *Building the Navy's Bases in World War Two, Vol. 1, Part 1*. Washington, D.C.: Government Printing Office.

Contents

Chapter 1
Greenland from Prehistory to 1731

Abstract Greenland is geographically a part of the North American continent but has remained politically aligned with Europe since the first European landing there in 978 CE. The indigenous settlement of Greenland arrived in fits and starts over thousands of years, as highly skilled and supremely adapted humans hunted and migrated across the Canadian high Arctic and eventually crossed to Greenland on kayaks or umiaks or on sledges pulled by equally well-adapted dogs. Humans are thought to have first arrived in Greenland some 4500 years ago and remained in the area of West Greenland for nearly two millennia. The European presence in Greenland was reestablished by the Dano-Norwegian missionary Hans Egede in 1721.

Keywords Greenland · Snæbjǫrn galti Hólmsteinsson · Hans Egede · Erik Thorvaldsson · Thorfinn Karlsefni Thórdarson · Paleo-Eskimo · Saqqaq · Independence I · Independence II · Douglas Clavering · William Scoresby, Jr. · Egil Knuth · Dorset · Thule · Icelandic sagas · Kingiqtorsoaq · Poul Nørlund · Hvalsey

Greenland is geographically a part of the North American continent but has remained politically aligned with Europe since the first European landing there in 978 CE by an Icelandic chieftain named Snæbjǫrn galti Hólmsteinsson. Galti's expedition was no less troubled than Ted Novak's nearly 1,000 years later during a global war. An Icelandic saga, now lost to time, recorded that Galti's fellow expeditioners disputed several points of order within the chieftain's proposed Greenland colony and, in true saga fashion, resolved the conflict by ending Galti's existence.

A far luckier chieftain arrived in southern Greenland a few years later by the name of Erik Thorvaldsson, much better known to history as *Eirík raude* or Eirík the Red. In the sagas, one is usually lucky or unlucky in life, such luck a direct residue of one's underlying character. This character, we are reminded in that still-quoted seventy-seventh stanza of Old Norse advice known as the *Hávamál* (Anon., n.d.), follows one into death:

P. J. Capelotti, *Lost Greenland*, Springer Polar Sciences,
https://doi.org/10.1007/978-3-032-27841-8_1

Kveg dør.
Slekt dør.
Vi dør også.
Jeg vet én ting som aldri dør.
Dømmen over hver eneste død.

Cattle die.
Kin die.
We die as well.
I know a thing that never dies.
Judgement over the dead.

In this, Eirík became the historical opposite of Snæbjǫrn Galti. It is true that Eirík was fleeing Iceland (as he had earlier fled his birthplace along the west coast of Norway), by some trouble he had created for himself. After a series of killings, Eirík was declared *utlagr*, literally 'outside the law,' meaning anyone could dispatch him with impunity. But since Eirík's own saga does survive, it leaves us with a strong portrait of adventures and misadventures with both the law and geography that Eirík turned to his advantage in both life and death.

Sailing west from Iceland, Eirík followed the keel waters of Snæbjǫrn Galti some 400 miles west until he landed upon a forbidding, mountainous shore of snow and glacial ice. A born opportunist, Eirík recognized a real estate opportunity when he saw one. He gave the uninhabited area the name 'Groenland' (Greenland), for, as his saga records without a trace of shame, Eirík "said people would be attracted there if it had a favourable name" (The Sagas of Icelanders (2000): 655).

Eirík found a wonderland that, if not exactly green, had at the least patches of the bright green mosses and scrub willow and birch heathlands that cling to the southern coastal areas of the island. Better still, for Eirík's colonizing expedition, the entire dramatic landscape of mountains and ice and snow seemingly held not a single other human. Years later, Thorfinn Karlsefni Thórdarson, descendent of the Danish king Ragnar Lodbrok, would follow in Eirík's son Leif's keel waters to the lands to the west that Leif had explored. There Karlsefni would most certainly encounter native Americans and ultimately be forced to abandon a Norse colony in the New World. But in the Greenland of the Saga period, no indigenous peoples troubled the early decades or even, apparently, the first centuries of Norse colonization.

The indigenous settlement of Greenland arrived in fits and starts over thousands of years, as highly skilled and supremely adapted humans hunted and migrated across the Canadian high Arctic and eventually crossed to Greenland on kayaks or *umiaks* or on sledges pulled by equally well-adapted dogs. The speed with which both humans and animals can travel across sea ice was shown nowhere so dramatically as in 2018, when a blue fox fitted with a tracking device left Svalbard in Arctic Norway on 26 March, reached Greenland in just 21 days, and continued onwards to Ellesmere Island in Nunavut, Canada where it arrived on 1 July. The dramatic and rapid crossing of some 3500 km (2000 miles) so stunned researchers that at first, they refused to believe it, instead hypothesizing that the radio collar had come loose and been taken on board a boat by humans (Fuglei & Tarroux, 2019; see also: Rourke 2019).

Humans are thought to have first arrived in Greenland some 4500 years ago and remained in the area of West Greenland for nearly two millennia. One population appears to have survived even as recently as 700 years ago. These Paleo-Eskimo peoples, known as the Saqqaq after a site where archaeological research has revealed much about them, utilized driftwood and finely crafted stone tools, and perhaps a form of the kayak as well, to exploit the High Arctic environment along the West Coast of Greenland. The absence of ready supplies of wood, however, would have made anything like a kayak or *umiak* boat a first-class rarity (Oswalt, 1999: 120).

These early migrations, independent of later migrations involving both native American and Inuit movements, appear to have originated in the far northeastern areas of Siberia, as linked genetically from a study of frozen Saqqaq-period remains uncovered near Disko Island (see, for example, Raghavan et al., 2014, Rasmussen et al., 2010).

The Saqqaq were joined, about a century after their arrival, by another Paleo-Eskimo population, now known as Independence I. Tracking musk ox and Arctic hare across the frozen landscape of the north coast of Greenland, these total populations are thought to have numbered no more than a handful of families. No one knows what these Paleo-Eskimo peoples called themselves, or precisely when these two migratory peoples ended their occupations of western and northern Greenland.

Both the Independence I and a later people known as Independence II lived north and south of what Robert Peary labeled as Independence Bay, when he reached the area with Norwegian explorer Eivind Astrup in 1892. It was the remarkable Danish archaeologist, sculptor, and lifelong explorer of Greenland Eigil Knuth, traveling across northeast Greenland before and after the Second World War by ship, sledge, and Tiger Moth biplane (before the war) and a U.S. war surplus PBY Catalina flying boat (after), who named the Independence cultures and split them into two separate occupations. Knuth himself believed Greenland had been his destiny since birth, his grandfather having financed Fridtjof Nansen's Greenland crossing in 1888 and later gifting Nansen's compass to his grandson upon his birth (Laursen, 1996).

Occupying the northern edge of the island, these High Arctic peoples, likely in their skin boats and sledges with dogs, would have possessed the kind of intimate geographic knowledge of the Greenland coast that had for so long eluded Europeans and Americans. In their ships, dog sledges, motor sledges, balloons, and airplanes, these later explorers struggled in vain to penetrate Greenland's coastal ice shield and inland ice sheet to learn, as but one example, just where Greenland ended and the Arctic Ocean began.

The first serious attempts to chart the coastline were carried out by the Yorkshire polymath William Scoresby, Jr., in his whaler *Baffin* in 1822 and a year later by Douglas Clavering in HMS *Griper*. It was Clavering, a young Royal Navy officer born at Holyrood House, grandson of both the commander-in-chief of India and the fifth Duke of Argyll, who was the first European to make contact with the indigenous people of Northeast Greenland since the arrival in the area of the Independence I families more than 4000 years earlier.

A century after Clavering, it became the preoccupation of Eigil Knuth's long life to understand how these prehistoric populations of Greenland had waxed and waned and had, through the various and complex migrations from north to south along the immense and immensely hazardous eastern and western coastlines, potentially influenced one another. In the mid-1930s, Knuth retraced his hero Nansen's steps across the inland ice and then constructed and settled into a house at Angmagssalik (Tasiilaq) in East Greenland to write up his adventures and create sculptures of local Greenlanders (Laursen, *op cit.*).

Knuth's fieldwork, ranging over thousands of miles of Greenland, established a cultural timeline from the first arrivals of the Saqqaq and their likely close relatives, the Independence I culture, and then the later Independence II culture as a possible expression of an early Dorset Inuit arrival in Greenland from Arctic Canada. All this would have accounted for a tremendous success, but Knuth also laid the groundwork for the study of the geology, zoology, and botany of what is now the Northeast Greenland National Park, the largest national park on the globe and the tenth largest protected area on the planet. At 375,000 square miles, the park is more than seventy-five time larger than Yellowstone National Park in northwest Wyoming.

Around 2500 years ago, a population of skilled sea hunters and dog drivers, possibly a relation of the Independence II people, made its way into northwestern Greenland, where they remained for more than a thousand years. These Dorset Paleo-Eskimo people created some of the most intricately beautiful bone carvings of seals and polar bears ever discovered from prehistory. Nomadic, reliant upon sea mammals both to survive and to light their characteristic soapstone lamps, the Dorset were supremely adapted through both patience and clothing to some of the most extreme cold conditions on the planet. This adaptation to extreme cold is what allowed the Dorset to survive at a time of cooler temperatures (see, for example, D'Andrea et al., 2011).

Archaeological evidence suggests that the Dorset had left Greenland a thousand years or more before the arrival of the Norse pagan Eiríк the Red and his colonists from Iceland in southwestern Greenland at the conclusion of the first millennium of the Christian era (see, for example, Jensen, 2006). The ancestors of the present-day indigenous Greenlandic population were the Thule Inuit.

Unlike the Dorset and the earlier Paleo-Eskimo cultures, the Thule survived in Greenland into the Dano-Norwegian colonization of Greenland in 1721 and so their culture has been studied and described for three hundred years. Rather than wait patiently for sea mammals out on the sea ice like the Dorset, the Thule employed dog sledges for rapid over ice transport in winter. In summer, a combination of one-person kayaks and large, cargo-carrying *umiaks* made from driftwood and sealskins made the inner seas around the coasts accessible. The Thule hunted caribou but most of their diet derived from sea mammals and local plants. Mummified remains of Thule Inuit buried in frozen coastal caves and rock crevasses reveal facial tattoos and, internally, lungs blackened by exposure to black soot particles from the burning of seal fat, as well as cancers from exposure to the Epstein-Barr virus (see, for example, Lynnerup, 2015).

The Thule arrived in northern Greenland from Alaska centuries after the arrival of Eiríc the Red and his compatriots. The Norse were clinging to a precarious existence in south and west Greenland when the Thule crossed from Arctic Canada over to what is now known as Thule, Greenland, around 1200 CE (Lynnerup *op cit.*).

By the time of the Thule migration into northwest Greenland, the Norse had been established in southern and western Greenland for the better part of two hundred years or more, as a minor trading outpost on the far edge of Europe. During this time, they had established numerous stone structures to house Augustinian monks and Benedictine nuns, "the first Christian monastic houses ever established in the North American continent," even a bishopric at Garðar (see, for example, Morgan, 2024).

This widespread newly Christianized area in southwest Greenland would likely have existed in complete ignorance of the Thule Inuit occupation far to the north along the vast northern and northwestern shore of Greenland. Just when the Thule moved south and interacted with the Norse, and whether this movement was triggered by climatic changes that rendered southern and western Greenland more favorable to a Thule way of life while simultaneously isolating Norse Greenland from its European lifelines, has been the subject of intensive scholarly study and debate for decades. This has become especially relevant, of course, as a warming world forces modern human populations to study the cultural survival strategies of earlier human settlements (see, for example, Dugmore et al., 2007).

What seems clear is that the image of the tall, almost all-powerful Norse explorer of the North Atlantic personified by Eiríc the Red and his son Leif had, through insular dwarfism throughout the medieval period, left the Norse in Greenland a much diminished people incapable of mounting a successful defense against incursions of Thule Inuit or raids by the new marauders of the North Atlantic in the 1400s, English pirates, slavers, and stock fishermen, along with Barbary fortune seekers.

But all that lay in the far future as the Norse expansion across the North Atlantic commenced between the early ninth and late tenth centuries of the Common Era. This dramatic expansion into and onto islands ranging from modest islets to the largest island in the world, involved adaptations both varied to local circumstances yet based upon a common maritime and ethical heritage. This common worldview derived in part from the development of Norse expertise in coastal farming, shipbuilding, and ship handling, combined with a profound if at times opaque literature as is contained in the medieval Icelandic sagas and the more than 160 gnomic oral traditions of the god Odin, 'the High One', contained in the *Hávamál* (see, for example, Jones (1986); Seaver (1996); Dugmore, *op cit.*).

The Norse expansion onto the southern and western coasts of Greenland took place in the middle of the Viking Age itself. Beginning with a raid on a church off the coast of northeast England in 793 of the Common Era, the time was marked by the gradual conversion of Odin-worshipping pagans to a monotheistic Christianity, from tribal raiding to regional assemblies to more centralized governments. These transitions can be seen in the architecture of Norse stave churches, where pagan dragons compete for prominence with Christian crosses.

The Norse were both explorers and traders across large swaths of Europe, the North Atlantic, and the Arctic. The Icelandic sagas, the record of many of the impulses of this era, coincide roughly in time with the Viking Age itself, from the settlement of Iceland to the defeat of Haraldr Sigurðarson at Stamford Bridge in September of 1066. No one embodied the complex human challenges of the era more than Eirík the Red himself, 'Eiríkur' being an old Norse name meaning 'powerful.'

Eirík was exiled from Norway at the moment in time when Icelandic society was forming itself into the world's first representative republic. Firmly a part of Europe, Iceland counted on Norway and England as its primary trading partners. The Icelandic sagas themselves were literal, neither mythical nor fantastic, and as such intended to inform as well as entertain, especially with regards to the critical elements that comprised success or failure in the North Atlantic: geography, history, and personal and community ethics. For half a millennium, explorers such as Eirík the Red held onto more precise geographic knowledge than all the geographers of Europe.

Eirík's Greenland colonizing expedition of 985–986 was a masterwork of administration, with fully fourteen of twenty-five vessels reaching the southern edges of the island. He was favored by a warmer climate that allowed the passage of shallow draft cargo vessels through the coastal ice, conditions which did not significantly deteriorate for some 350 years. Those warmer waters had also served to keep the High Arctic Dorset and the later Thule Inuit within their own favored and much colder adaptations along the northern, northwestern, and northeastern coastlines of Greenland.

Eirík and his compatriots thus found an uninhabited otherworld where deep Greenlandic fjords teemed with sea life. There they could establish, one to a family or clan, their own private expanses of coastal grasses and scrublands which, added to the harvest from the sea, contributed to the establishment and maintenance of a Norse way of life. And there the pagan Eirík erected a small chapel for his converted wife—and the mother of Leif the Lucky—Thjodhild Jorundardóttir.

The arrival of the Norse in Greenland marked the most lasting European colony on the western edge of the sea road of the North Atlantic. Across five centuries, the descendants of Eirík the Red and his fellow colonists maintained a Norse outpost that now seems to appear as a distant specter in Icelandic and other medieval European accounts of northern peoples. Eventually, Eirík's minute democracy with its small religious observances grew into an annual assembly that was held at Garðar (Igaliko). A Greenlandic bishopric followed in 1126. The foundation walls of the Garðar cathedral can be seen to this day, as can the much more spectacular ruins of the Hvalsey Church near present-day Qaqortoq (Fig. 1.1).

The largest attempt to settle these already settled lands to the west of Greenland was made around the time of the death of Eirík the Red in 1003. An Icelandic merchantman by the name of Thorfinn Karlsefni Thórdarson arrived in Greenland and met and married Gudrid Thorbjarnardóttir, the widow of Eirík the Red's youngest son Thorstein. The Greenlandic sagas have Thorfinn and Gudrid at the head of an expedition of some three or more ships and between 60 and 140 crew.

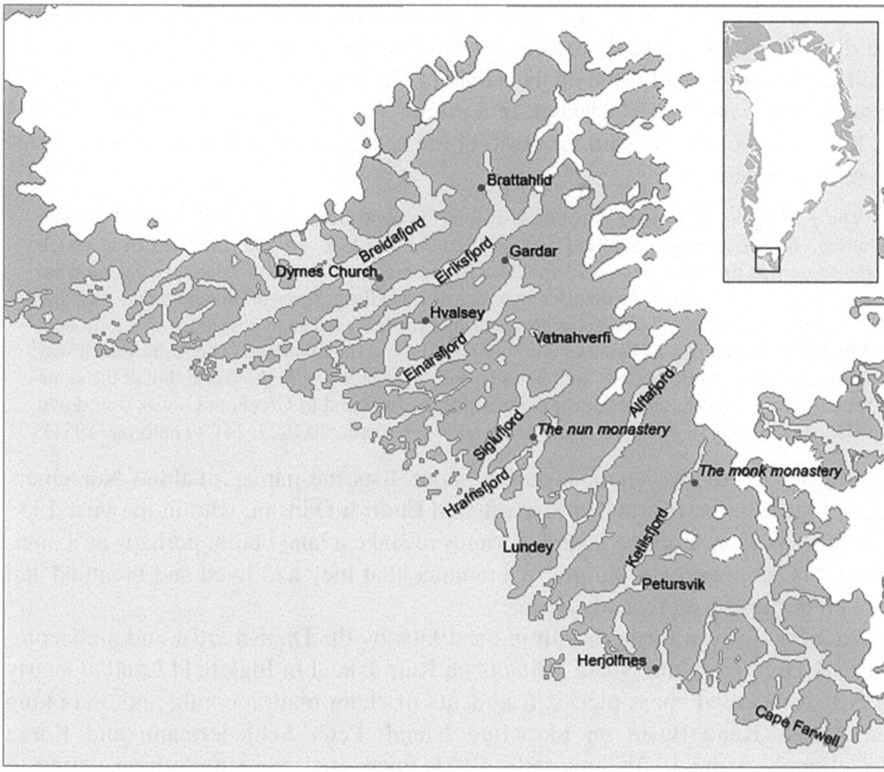

Fig. 1.1 From its earliest years, the colony seems to have not just survived but possessed the excess energy to explore the seas even further to the west and the north. Attempts by the children of the Norse colonists—Bjarni Herjólfsson, son of Bárdi Thorgerdr Herjólfsson who established the southernmost Norse outpost at Herjolfnes; and of course, Eirík and Thjodhild's son Leif Eiríksson—either spotted or stepped ashore at lands to the west now thought to be Baffin Island, Labrador, and Newfoundland (Source: Masae via Wikipedia Creative Commons)

By the accounts of the sagas, this expedition can be seen as a dramatic first attempt to establish a European colony on what became North America, even seeing the birth of the first European on American soil, Leif's nephew Snorri. Read another way, Karsefni's voyage involved a succession of calamities. One of these triggered the determined and successful opposition of native Americans to a European foothold upon indigenous lands and led to the death in battle of Leif's brother Thorvald. Another revealed the criminal insanity of Leif's sister Freydis. The failed colony played out in the best traditions of Wagnerian opera, with the curses of ambition, an insatiable striving towards a near-holy grail, and fame and love beyond death.

With such a rich medieval stew of its founding, one that was eventually memorialized in the two surviving sagas of the Greenlanders, it is small wonder that the long history of the Norse Greenland colony after the generations of Eirík and Leif, Gudrid and Snorri, would find itself relegated to an historical footnote. Yet the hints we have of the life of the colony for at least three hundred years indicates a continued dynamism. In 1824, a Greenlandic Inuit named Pelimut (Philemon) located a small (10 cm) inscribed stone on the island of Kingiqtorsoaq ("towering mountain") about 10 miles north of Upernavik in northern West Greenland, in latitude 72° 58' N., longitude 56° 14' W, and far north of the established Norse colony settlements in western Greenland.

> "The Kingigtorssuaq stone is a blackish-gray or dark-green quartz slate, with a smooth surface for the inscription, found on stony ground beside a very old ruined cairn in which the stone was probably once placed. Close by, at the top of the same island, two somewhat smaller cairns are found, so that the three cairns form the corners of an equilateral triangle.
>
> "These cairns may have given the bearings of the permanent residence of the people who had erected them as being on another island or on the mainland. The same year it was found, the small runic stone—undoubtedly one of the smallest in the world, but at the same time the most significant of all runic inscriptions yet found in Greenland—was sent down to Copenhagen by W. A. Graah on his first voyage to Greenland (1823–24)" (Thalbitzer 1951).

The translation of the small, inscribed stone lists the names of three Norsemen, Erling Sighvatsson, Bjarni Thordarson, and Endridi Oddson, who in the year 1333 raised three cairns atop the island, perhaps to stake a land claim, perhaps as a navigational beacon, perhaps simply to announce that they had lived and breathed and were embarked on a great adventure.

Excavations even further north in the 1930s by the Danish artist and archaeologist Erik Holtved found Norse artifacts on Ruin Island in Inglefield Land at nearly 80°N that included chess pieces, fragments of chain mail, a comb, and a cooking pot. Across Kane Basin on Skraeling Island, Peter Schledermann and Karen McCullough in the 1970s and early 1980s uncovered woolen clothing dating to approximately 1280 CE, smelted copper and iron, pieces of chain mail and ship rivets (McCullough, 1989). Since the Thule Inuit are thought to have occupied the island and the Kane Basin area around this same time, taken with the Kingiqtorsoaq inscription, these researches suggest that Norsemen were exploring and trading into the High Arctic for hundreds of years.

Icelandic annals thought to have belonged to the bishop of Skálholt in southern Iceland record a similar encounter from around the same period. A Norse ship with a crew of seventeen or eighteen had visited Markland (likely modern-day Labrador), no doubt in search of the ready timber unavailable in Greenland, and on their return to the Norse colony had been blown off course and landed instead in Iceland (see, for example, Reeves et al., 1906).

The year of the Kingiqtorsoaq rune is the subject of some speculation, but not its authenticity. Like the 1347 voyage to the forests of Labrador, it is a testament to a still-exploring society, one motivated by the competitions of Norwegian kings and the bishops of Rome over tithes and trading throughout the geography of the North Atlantic bishoprics (see, for example, Seaver, *op cit.*: 76–81). It was not yet the lonely outpost it became in later years: cloistered both literally and figuratively from the European church and royalty that forgot about it. Never more than about 3000 souls, the colony was necessarily scattered along deep and remote fjords and therefore poorly situated against pirates and plagues, or the skilled Arctic hunters who appeared from the north as the world grew warmer and the trading and shipping lanes that closed as the world turned colder.

The year of the visit from Greenland to Labrador is particularly poignant, as at the same moment, Europe was about to suffer the loss of one-third of its population to the pandemic of the bubonic plague. The plague would both devastate Europe and contribute to the sudden isolation of the Greenland Norse colony from the life of Europe. Long established trading routes were suddenly abandoned for lack of sailors and ships, and the sailing route to Greenland, never a simple proposition in the best of weather, was suddenly not worth the risk.

In addition to this catastrophe, the once-reliable Greenland walrus ivory trade through the port of Bergen suffered from new imports of cheaper ivory from Africa and Asia. The last recorded ship sailing from Norway to Greenland took place in 1367. Simultaneously, and perhaps because of the plague reducing its own population either directly or indirectly, and combined with the movement of the Thule Inuit southwards—which according to the Icelandic Annals of 1379 included a disastrous event when eighteen Norse were killed and two boys captured by Inuits—the Norse abandoned the northern farms that they had established inland from the modern Greenland capital at Nuuk.

In the early 1920s, the Danish cultural researcher Poul Nørlund was given leadership of an archaeological expedition to investigate the Norse settlement of Herjólfsnes, that port at the southern tip of Greenland that would have served as the first port of call for ships visiting from Europe and the final port of departure for traders leaving Greenland for home. There, despite egregious vandalism of the site by a visiting ship's crew in the mid-1800s, Nørlund located Christian burials wrapped in well-preserved Norse woolen clothing.

Nørlund's work, which soon led to the creation of the Commission for the Scientific Investigation of Greenland, seemingly revealed that the Norse colony eventually reestablished regular contacts with Europe following the long decades of the plague. The Herjólfsnes clothing bears stylistic resemblances to European clothing dating to the late 1400s. Even in its final years, the Norse colony apparently clung to its European religion and its European clothing.

Soon, however, Norse Greenland would vanish from European memory as well. The last recorded Norse wedding took place at Hvalsey church in 1408; the last recorded ship reached Iceland from Greenland just 2 years later. The Vatican continued, after the death of the incumbent Bishop of Garðar around 1410, to appoint new bishops to Greenland, but none seems to have bestirred themselves to actually make the journey to their new post.

The pope himself in 1448 complained that Greenland had been without a bishop for 30 years, but still none of the repeated appointees seemed eager to test to icy straits around southern Greenland. A similar letter was written from Rome in 1519, with the same result, by which time the medieval Norse church in Greenland had been without a living church authority in over a century (Reeves, et al., *op cit.*). After 1520, Norse Greenland seemingly vanished behind a veil of half-remembered Icelandic texts and pack ice.

By the time of the voyages to Newfoundland of the Italian born, English chartered Giovanni Caboto (John Cabot) in the late 1400s, it is an open question whether knowledge of Norse Greenland lingered on with Cabot and his crew, or if in fact any interactions took place between medieval Norse Greenlanders and Renaissance buccaneers. By Cabot's time, the Viking Age had been over for more than four centuries, the near-mythical marauding sea raiders that led Irish monks to pray for bitter storms to keep Norse ships away a distant memory. It had been at least two hundred years since the oral legends of the Norse had been written down and then edited and blended, lost and found and edited again, to suit the politics and the gods of newer generations.

Visits by fifteenth century English ships in search of resources from fish and up to and including human slaves are recorded in Icelandic history. It is no great stretch that these same forces reached and exploited nearby Greenland and its remaining few beleaguered colony survivors, or in fact that the surviving Greenlanders willingly joined the buccaneers in their various searches for new lands and homes in North America. In this way they would—albeit five centuries later—have achieved Karlsefni's vision for a Vínland colony in the New World.

When Hans Poulsen Egede was born in Harstad in northern Norway in January of 1686, it is likely that the last Norseman in Greenland had been dead for two hundred years or more. From a family of Lutheran pastors, Egede studied at the University of Copenhagen before being ordained in 1707 and assigned to a parish in the Lofoten islands of northern Norway. There Egede heard tell of Norse settlements in Greenland, apparently long lost to the faith, and determined to visit for himself.

Egede raised enough capital from Bergen merchants to make an exploratory voyage in the summer of 1721 that included his wife and four children and another forty hopeful colonists. Landing in southwestern Greenland, Egede searched in vain for his fellow Norsemen and encountered instead the very much living descendants of the Thule Inuit. He began at once to learn the Inuit language, and his charter from the Bergen Greenland Company had given him wide latitude over the presumed land and riches to be found on the land and seas in and around Greenland.

By the end of 1721, many of Egede's hopeful colonists were dead of scurvy or had decided to take the first ship home. But Egede and his family remained, and 2 years later uncovered the ruins of the long-lost Norse colony. Egede helped establish the port town of Good Hope (Godthåb), present-day Nuuk, in 1728, and a pious bronze statue of Egede stands watch over both Nuuk and outside Frederick's Church in Copenhagen. Some 735 years after the similar journeys of the Norwegian rogue Eirík the Red from Iceland, the Norwegian priest Hans Egede had established a Nordic colony in southern Greenland once more.

References

Anon. (n.d.). Translation of Havamal passage by the author.

D'Andrea, W. J., et al. (2011). Abrupt Holocene climate change as an important factor for human migration in West Greenland. *Proceedings of the National Academy of Sciences of the United States of America (PNAS), 108*(24), 9765–9769.

Dugmore, A. J., et al. (2007). Norse Greenland settlement: Reflections on climate change, trade, and the contrasting fates of human settlements in the North Atlantic Islands. *Arctic Anthropology, 44*(1), 12–36.

Fuglei, E., & Tarroux, A. (2019). Arctic fox dispersal from Svalbard to Canada: One female's long run across sea ice. *Polar Research, 2019*(38), 3512. https://doi.org/10.33265/polar.v38.3512

Jensen, J. F. (2006). The stone age of Qeqertarsuup Tunua (Disko Bugt): A regional analysis of the Saqqaq and Dorset cultures of Central West Greenland. *Meddelelser om Grønland, Man & Society, 31*, 1–272.

Jones, G. (1986). *The Norse Atlantic Saga* (2nd ed.). Oxford University Press.

Laursen, D. (1996). Eigil Greve Knuth 1903-1996. *Arctic, 49*(4), 401–403.

Lynnerup, N. (2015). The Thule Inuit mummies from Greenland. *The Anatomical Record, 2015*(298), 1001–1006.

McCullough, K. M. (1989). *The ruin islanders: Early Thule culture pioneers in the Eastern High Arctic*. Canadian Museum of Civilization/Archaeological Survey of Canada. Mercury Series Paper 141.

Morgan, R. J. (2024). *Monastic orders in medieval Greenland*. Online at: Retrieved September 15, 2025, from https://rebeccajanemorgan.medium.com/monastic-orders-in-medieval-greenland-ebb3d75feba2

Oswalt, W. H. (1999). *Eskimos and explorers* (2nd ed.). Lincoln: University of Nebraska Press.

Raghavan, M., et al. (2014). The genetic prehistory of the New World Arctic. *Science, 345*, 6200). Online access at:. https://doi.org/10.1126/science.1255832

Rasmussen, M., et al. (2010). Ancient human genome sequence of an extinct Palaeo-Eskimo. *Nature, 463*, 7282). Online access at. https://doi.org/10.1038/nature08835

Reeves, A. M., Beamish, N. L., & Anderson, R. B. (1906). *The Norse discovery of America*. The Norroena Society.

Seaver, K. A. (1996). *The frozen Echo: Greenland and the exploration of North America ca a.D. 1000–1500*. Stanford University Press.

The Sagas of Icelanders. (2000). *Introduction by Jane Smiley*. Penguin.

William, W. T. (1951). "Two runic stones, from Greenland and Minnesota." Washington, D.C.: Smithsonian Institution.

Chapter 2
Lost Greenland: Benjamin Franklin, Thomas Jefferson, Samuel Morton, and the Arrival of Greenlanders in Philadelphia

Abstract The first American to express a deep interest in Greenland and its people was Benjamin Franklin, who actually met with Greenlanders in Philadelphia during a tour organized by the Moravian Church. Franklin's interest was followed by Thomas Jefferson as he sought the origins of language. In the early nineteenth century, Samuel Morton sought to use the myth of the Lost Colony of Greenland to influence both racial attitudes and the next generation of explorers to Greenland, beginning with Elisha Kent Kane.

Keywords Greenland · Greenlandic Inuit · Norse · Ívar Bárdarson · Greenlandic Inuit · Philadelphia · Northwest Passage · New Sweden · Moravians · Racism · *Crania Americana*

> *Most readers are aware that colonies of Scandinavians and Icelanders peopled Greenland in the Middle Ages. Since the fourteenth century, however, nothing has been heard of them, and they were supposed to have been blocked up and destroyed by the accumulating ice, whence the name of Lost Greenland. In 1829 the Danish government sent Captain Graah to explore these icy solitudes, and to ascertain at least the locality of the lost colony.*
> —Samuel Morton, *Crania Americana* (1839)

The American interest in Greenland did not begin with businessman and former U.S. ambassador to Austria Ronald Lauder's suggestion to Donald Trump in 2019 that the U.S. purchase Greenland from Denmark (see Baker & Glasser, 2022; Chitirala & Ryan, 2025). American history in Greenland is generally traced to Elisha Kent Kane's participation in the many searches for the lost expedition of British explorer Sir John Franklin in the 1850s. Yet the American fascination with the island goes back much further, to more than a quarter century before the founding of the United States itself and, as with Kane, it centers in the city of Philadelphia.

While pieces of this history have often been told—and many have entered the American historical memory—the sweep of that history dates all the way to the anthropological fascinations of the scientists and statesmen amongst the Founding Fathers, specifically Benjamin Franklin and Thomas Jefferson. The quote above from Philadelphia physician and phrenologist Samuel Morton is off by a century

P. J. Capelotti, *Lost Greenland*, Springer Polar Sciences,
https://doi.org/10.1007/978-3-032-27841-8_2

when describing the loss of contact with the Greenland colony, and modern archaeological studies like those at Herjofnes suggest that the colony may well have survived into the early sixteenth century. The Danish king Christian IV himself sent three expeditions in search of Greenland between 1605 and 1607.

What is important here is not the actual life and ultimate disappearance of the Norse presence in Greenland, but rather, as University of Copenhagen's Robert Rix puts it, the question of "how Greenland's vanished settlers became a cultural memory…"

> Enduring narratives of Greenland were established based on a mélange of garbled facts, misunderstandings, and sometimes runaway imagination. When it came to conceptions of precious resources, Greenland and its settlements found a place in the discourse of Western imperialism… how the 'lost colony′ lived on in the Western imagination… (and) memorialized a colonial past… (Rix, 2023: 2–3).

Of course, the Greenland colony was never truly lost, in that almost anyone from 1400–1700 with an interest in what resources might be found there had access to a report by a medieval Norwegian cleric's report. Ívar Bárdarson spent twenty years in Greenland and his report from the mid-1300s was available in translation throughout the North Atlantic world even into the early 1700s (Rix *op cit.*: 29–30). Bárdarson's. descriptions of church property, fishing and whaling grounds, walrus ivory, therapeutic hot springs, all these—later embellished with the usual claims of mountains full of silver and gold and an imagined geographic connection between Greenland and the fabulous riches of India, or more directly to North America—would never have escaped the attention of anyone planning a serious venture after trade or plunder in rich North Atlantic waters.

A familiar thread woven into the eventual expeditions sent in search of the supposed lost colony was the ancient trope of peaceful Christians under continuous siege by pagans. "The idea that savages would attack and massacre peaceful European settlers is found soon after communication with Greenland was lost" (Rix *op cit.*: 49). When Hans Egede published his *A Description of Greenland* in 1741, he had whittled the many possibilities for the disappearance of the Greenland Norse down to just one: predations by Greenlandic Inuit (Rix *op cit.*: 57).

Whereas much of this contemporary speculative history of Greenland centered on the notion that there was a European, Christian colony in the High Arctic that was somehow, and some moment, 'lost,' either though the encroachments of the ice or those of humans considered very much the lesser of these God-fearing Europeans, both Franklin and Jefferson approached Greenland as leading lights of the American Enlightenment. In modern terms, Franklin's interest was very much that of a cultural and economic anthropologist, while Jefferson's approach relied on archaeology and linguistics. Each in their way employed inductive generalization to their research, gathering specific evidence on their interests before proceeding to more general probabilities.

Egede's expeditions to Greenland in the 1720s, and his conclusions from them, took place at the precise moment a young compositor from America was working in a print shop in London. It is unlikely that Egede's work escaped the notice of the first American to become fascinated with the island, the voracious reader and largely self-taught Benjamin Franklin. The year that Franklin returned from London to Philadelphia in 1726, Egede was building the first chapel at Godthåb. Egede's *Det gamle Grønlands nye Perlustration* appeared in 1729 and was translated into several languages including English.

Two years later, Franklin established the first library in Philadelphia and Egede's volume can be found there. Franklin's Library Company of Philadelphia was founded just 10 years after Denmark colonized Greenland, and its earliest surviving catalog, from 1741, includes volumes on exploration, geography, and history. The collection today catalogs Isaac de La Peyrère's 1647 *Relation du Groenland* and Pierre Martin de La Martinière's 1674 *A new voyage into the north countries…* which included "reflexions (sic) upon an error in our geographers abou (sic) the situation (sic) of Greenland…" As Rix notes, La Peyrère's volume "achieved Europe-wide circulation during the seventeenth and eighteenth centuries" (Rix, *op cit.*: 42).

It is unlikely these volumes escaped Franklin's wide span of attention, especially since La Peyrère's popular account of Greenland was translated into English in 1732. Franklin came of age as well during raging debates over the existence and plausibility of a Northwest Passage to Asia. As a British colonial citizen, Franklin's interests in geography and trade led naturally towards the study of some of the fundamental geographic mysteries of his age, for example the nature and existence of sea passages to Asia across the top of the world.

Franklin's Library Company expanded during the same decade as the Danish-born Russian cartographer Vitus Bering's Great Northern Expedition, which led to the European discoveries of Alaska and the Aleutian Islands. "Although Bering and his associates did not locate the Northwest Passage, they coasted northwest North America (and) led to the Russian fur trade in Alaska and the English fear that Russia might succeed in navigating the North-west Passage first" (Solis-Cohen, 1957: 17).

Franklin's first direct mention of Greenland in his writings appeared in his *Poor Richard's Almanack*, an annual compilation of Franklin's wit and wisdom written pseudonymously under the name of Richard Saunders. Amongst the journal's predictions for the coming year of 1739 was: "a plentiful Year of all manner of good Things, to those who have enough; but the Orange Trees in Greenland will go near to fare the worse for the Cold" (Franklin, 1739).

Franklin was among the leading figures in the movement to free America from British rule, but he was so much more than that. His successful journalism and newspaper and almanac publishing led him to become an influential figure in the city of Philadelphia, helping to shape it as the center of American politics and science. A prolific writer, musician, and chess player, he established the American Philosophical Society in 1743 to advance the study of both the humanities and natural sciences. As a prominent figure in adapting the European Age of Enlightenment to America, he conducted experiments into the nature of electricity, studies of American demographics, observations of the currents that circulate within the North

Atlantic Ocean, and numerous other inquiries in physics, meteorology, aeronautics, and refrigeration.

In addition to his scientific acumen, Franklin was regarded as a master diplomat during tenures as minister to both France and Sweden. In the latter post, he benefited from the history associated with the Delaware River valley's New Sweden colony, in existence from 1638 to 1655. Villages alongside the Schuylkill River that now form the core of Philadelphia itself were named Nya Korsholm and Nya Vasa while just a few miles downriver was located the settlement of Nya Göteborg. In the fall of 1748, Franklin was visited by Swedish botanist and Linnaeus disciple Pehr Kalm. Kalm called Franklin "my very special friend," and described their excited conversation on the subject of Greenland.

> [A sea captain reported on the high summer temperatures north of the seventieth degree of latitude and concluded that the summer heat at the pole must be even higher because the sun shone so long without setting.] The same account with similar consequences Mr. Franklin had heard from sea captains in Boston, who had sailed to the most northern parts of this hemisphere. But still more astonishing is the account he got from Captain Henry Atkins, who still lives at Boston. He had for some time been fishing along the coasts of New England, but not catching as much as he wished, he sailed north as far as Greenland. At last he went so far that he discovered people who had never seen Europeans before (and what is more astonishing) who had no idea of the use of fire, which they had never employed, and if they had known it, they could have made no use of their knowledge, as there were no trees in the country. But they ate the birds and fish raw. Captain Atkins got some very rare skins in exchange for some trifles (see Benson, 1937 and Larsen, 1945).

As Franklin's wealth and influence grew, so did his library. In addition to books and manuscripts it eventually contained anthropological artifacts as well. Prominent among these were parkas and tools of the people and culture from the area around Smith Sound, who in the centuries since have variously been referred to as Arctic Highlanders, Polar Eskimo, Greenlandic Inuit, and the Inughuit. These items from Greenland likely arrived with a visit to Philadelphia of three native Greenlanders in 1749. That summer, two boys and a young woman, Christian converts of Moravian missionaries in the Arctic named 'John, Matthew, and Judith Issek' arrived in the city on board the Moravian mission ship *Irene* at the conclusion of a tour that included previous stops in England, Holland, and Germany. Two other Greenlanders had died during the earlier legs of this religious promotional tour.

In 1896, the Moravian Historical Society drew upon its own archives as well as newspapers published in Philadelphia and New York between the years of 1748 and 1758, to assemble an account of the voyages of the *Irene*, a ship in the society's service for nine years in the mid-eighteenth century (Jordan, 1896). During that time, *Irene* made no fewer than two dozen Atlantic crossings and one voyage to Greenland. In mid-June, 1749, *Irene* docked in Philadelphia, prior to the Greenlanders visit to Bethlehem, Pennsylvania, then as now the center of Moravian worship in the Americas.

Intercultural acceptance was a professed element of Moravianism and, in Bethlehem, the exchange of cultures truly began. "Around the Greenlanders and the Arawaks, more than thirty American Indians took their places: Delaware, Mahikan (sic), and Wamponoag (sic). The missionaries and members of the congregations filled the remaining space surrounding the non-Europeans. As everyone sang in their own language, the (community) diary notes: "This was something very special and heavenly—a concert without parallel." For the Bethlehem Moravians the biblical promise of the gathering of people from all nations had become a reality. During the evening service (Matthaeus) Stach spoke Greenlandic as he bade the congregation farewell" (Moravian Archives, 2011).

"The following day the Greenlanders set out for Philadelphia, where they visited the Governor, and from thence proceeded to New York." A correspondent in Philadelphia (very likely Franklin himself), writing under date of June 15th, to the *New York Gazette and Post Boy* states:

> "The Beginning of this Month three Natives of Greenland, two young Men and a young Woman, converted to the Christian Religion by the Moravian Missionaries in that Country, were on a Visit in this City to the Brethren here. They left their own Country about two years ago in a Ship belonging to that Society (which had carried a ready framed Church to be erected there, Greenland affording no Wood for Building; This vessel sailed from this Port a few days ago for Davis's Straight, with some of the Greenlanders on board) and have visited the Brethren in several Parts of Europe, as England, Holland and Germany. They were clad in Seal Skins with the Hair on, after the manner of their own Country, their Eyes and Hair black, like our Indians, but their Complexion somewhat lighter. The Moravians it seems, have a Mission at Berbice near Surinam, and two Indian Converts from there with these Greenlanders, met lately at Bethlehem in this Province with some of the Delaware Indians, and some of the Mohickons, converts also of the Moravians; and tho' their Native Lands are vastly so remote, as the Latitude 54 and 65 North, yet what they observed of each others Hair Eyes and Complexion, convinced them that they were all of the same Race. They could find, however, no kind of Similitude in their several Languages" (Jordan, *op cit.*: 69–70).

Franklin lived and moved with ease amongst both living Native Americans and an almost bewildering array of religious sects in Colonial America, seemingly with much more curiosity than prejudice. And the Moravian conference in Bethlehem brought together not only representatives from these Pennsylvania locals but peoples from as far afield as Greenland in the High Arctic to indigenous Arawaks from as far as the Moravian mission in what was then the Dutch colony of Berbice and is now the independent Co-Cooperative Republic of Guyana. The Moravians had been operating in Greenland since 1733, barely a decade after their own founding as a sect, filling a void created as Hans Egede's fledgling Bergen Greenland Company colony teetered on the brink of collapse.

In pre-Revolutionary America, Franklin was an astute observer of shifting rivalries and alliances, and that between the Moravians and the Delaware was no different.

That interest went two ways. As a recent study noted:

> Prior to the American Revolution, Pennsylvania's unique political dynamics allowed the Moravians to garner a degree of political influence that eluded them in other colonies and convinced Delaware leaders that the Brethren could be a valuable political ally in their quest to secure a homeland and protect their people. Despite the Moravians' marginal social position, they successfully developed connections within both Pennsylvania's Quaker and

> Proprietary factions. Moravian office-holders including William Edmonds, Timothy Horsfield, and William Henry cultivated associations with prominent leaders such as Governor Robert Morris and Benjamin Franklin and used their influence to address the concerns of those living in and around Moravian communities (Miller 2017).

At the conclusion of their Pennsylvania visit, the converted Greenlanders, along with missionaries Christian David and Matthaeus and Rosina Stach, were safely returned on 30 July to the Moravian mission of Neu-Herrnhut (in present-day Nuuk), Greenland, after a voyage of 4 weeks. As a student of at the intersection of culture, geography, and an honest profit, Franklin no doubt took note of how the Moravians managed to finance their expanding network of missions in the North Atlantic, now centered just 60 miles north of Philadelphia in the newly-established Bethlehem.

Within 4 years of the visit of the Greenlanders, Franklin had formed yet another company, this one to subscribe supporters in the colonies for two expeditions to open trade with Asia via a Northwest Passage. If no such passage could be discovered to the west of Hudson's Bay, then the expedition was tasked to evaluate the coast of Labrador, opposite Greenland, "for the improvement of trade, fishing, and whaling" (Solis-Cohen *op cit.*: 19).

> (T)hese two ventures to locate the Northwest Passage to the Indies went unmentioned in Franklin's autobiography. His own omission may account for the fact that few writers have mentioned his organization of the Northwest Company and the Philadelphia ventures in arctic exploration which he helped to promote. Franklin's attempt to open trade with Labrador, and possibly with the East Indies, deserves attention. It is certainly of significance that eighteenth century Philadelphia, as the second city of the British Empire, shared in the impulse for exploration and discovery which characterized the great age of Captain Cook and Captain Bligh; and that its leading intellectual figure and its leading merchants gave support to the enterprise, with the blessing of the Proprietor, Thomas Penn (Solis-Cohen 1943).

While ultimately unsuccessful, Franklin pursuit of both knowledge and profit in the Arctic likewise engaged the interest of another founder. Franklin's interests in linking race with language, using Greenlanders as a model, was a theme his revolutionary compatriot Thomas Jefferson would also explore when Jefferson wrote *Notes on the State of Virginia.*

In 1775, Jefferson arrived in Philadelphia, soon to head a committee that included Franklin, to draft the Declaration of Independence. One component of that declaration echoed similar exhortations against native populations reaching back to the 'Skrælings' of the Vínland sagas and continuing to the arrival of the Thule Inuit in the areas of the Greenland Norse. The King of England was cited for the Colonists' belief that he had "endeavoured to bring on the inhabitants of our frontiers, the merciless Indian Savages, whose known rule of warfare, is an undistinguished destruction of all ages, sexes and conditions."

The declaration guaranteed war with England, and Jefferson returned to Virginia to become governor of his home state at perhaps its worst moment. Turned out of office amidst the British army invasion of Virginia in 1781, Jefferson retreated west. His *Notes on the State of Virginia* was written during the early 1780s during this particularly fraught period in his life, as he escaped pursuit by the British and suffered the deaths of his wife and two infant daughters.

Despite it all, Jefferson nevertheless found time to ponder the origins of the Greenlanders. In the case of the Greenlanders, he anticipated the much discussed (and much critiqued) Solutrean Hypothesis whereby Greenland (along with Iceland and Labrador) had seen European sea rovers visit in the far past.

> Great question has arisen from whence came those aboriginals of America? Discoveries, long ago made, were sufficient to show that the passage from Europe to America was always practicable, even to the imperfect navigation of ancient times. In going from Norway to Iceland, from Iceland to Groenland, from Groenland to Labrador, the first traject is the widest: and this having been practised from the earliest times of which we have any account of the part of the earth, it is not difficult to suppose that the subsequent trajects may have been sometimes passed. Again, the late discoveries of Captain Cook, coasting from Kamschatka to California, have proved that if the two continents of Asia and America be separated at all, it is only by a narrow strait. So that from this side also, inhabitants may have passed into America: and the resemblance between the Indians of America and the eastern inhabitants of Asia, would induce us to conjecture, that the former are the descendants of the latter, or the latter of the former: excepting indeed the Eskimaux, who, from the same circumstances of resemblance, and from identity of language, must be derived from the Greenlanders, and these probably from some of the northern parts of the old continent. A knowledge of their several languages would be the most certain evidence of their derivation which could be produced. In fact, it is the best proof of the affinity of nations which ever can be referred to. How many ages have elapsed since the English, the Dutch, the Germans, the Swiss, the Norwegians, Danes and Swedes have separated from their common stock? Yet how many more must elapse before the proofs of their common origin, which exist in their several languages, will disappear? It is to be lamented then, very much to be lamented, that we have suffered so many of the Indian tribes already to extinguish, without our having previously collected and deposited in the records of literature, the general rudiments at least of the languages they spoke. Were vocabularies formed of all the languages spoken in North and South America, preserving their appellations of the most common objects in nature, of those which must be present to every nation barbarous or civilized, with the inflections of their nouns and verbs, their principles of regimen and concord, and these deposited in all the public libraries, it would furnish opportunities to those skilled in the languages of the old world to compare them with these, now, or at any future time, and hence to construct the best evidence of the derivation of this part of the human race…
>
> A separation into dialects may be the work of a few ages only, but for two dialects to recede from one another till they have lost all vestiges of their common origin, must require an immense course of time; perhaps not less than many people give to the age of the earth. A greater number of those radical changes of language having taken place among the red men of America, proves them of greater antiquity than those of Asia. (Jefferson, 1832; See also: Jefferson 1785 (1975); White 2015).

As in so many things, Jefferson was decades if not centuries ahead of other thinkers in developing a variety of hypotheses to account for the racial, cultural, and linguistic dispersals of human groups. The Solutrean hypothesis, using some highly suggestive archaeological evidence from the Delmarva peninsula, from a stone axe dredged up off the coast of Virginia in 1970, as well as haplotype evidence, argues the case for Paleolithic Europeans of some 20,000 years ago migrating along the ice front of the last glacial maximum, to reach and spread across North America.

Critics of the notion are numerous and include oceanographers skeptical that the ice front maximum and the archaeological data match within even thousands of years, archaeologists who see little evidence for Solutrean Period seafaring abilities, and geneticists who see consistent evidence for human migration from Asia, not Europe. Nevertheless, such a theory was bound to appeal to racists, with one observer noting that "it is very popular among white supremacists, who are fond of the idea of the original settlers of the continent being of European rather than Asian heritage." (White *op cit.*).

Such racial formulations were hardly new in discussions of the settlement of the Americas, but they became hyper-charged in the nineteenth century as the founding generation of liberal social thinkers and empirical scientists like Franklin and Jefferson left the scene. Franklin and Jefferson had dared to imagine a new order of reason and science in human affairs yet were ultimately eclipsed by Southern landowners who sought and succeeded in the creation of their own vision of a New World Order in America, one of native American displacement, the expansion of human bondage, and finally the attempted destruction of the very Union that had opened the way for their ambitions in the first place. As Roger Kennedy has written, Southern plantation owners in the space between Jefferson's Louisiana Purchase in 1803 and his death 23 years later "swept aside the subsistence economy of the American Indians of the South, sustained the trans-Atlantic shipment of labor from Africa, and filled the Southland with African slaves while its independent farmers were thrust to the sidelines" (Kennedy, 2003).

Such was the established moral architecture of the American nation upon the arrival of perhaps the greatest American proponent of the imagined connections between racial inheritance and social outcomes based upon the morphology of human skulls. Samuel Morton was both founder and chief faculty member of the Pennsylvania Medical College in Philadelphia. Morton founded the college in the same year that he published his massive study *Crania Americana: A comparative view of the Skulls of Various Aboriginal Nations* (Morton, 1839).

In this influential work, published simultaneously in Philadelphia and London, Morton's scope was nothing less than the entirety of human cultural and physical anthropology. In the present case, while discussing the 'Varieties of the Human Species,' Morton offered such considered cultural analysis as that "in the moral scale" the natives of "the icy shores of the great island of Greenland .... rank extremely low.... Their mental faculties, from infancy to old age, present a continued childhood: they reach a certain limit and expand no farther" (Morton *op cit.*: 54).

Such overt expressions of racism were hardly limited to the Greenlanders. Morton's discussion of other members of the 'Polar Family' encapsulates his method of juxtaposing physical characteristics with enumerated cultural and moral attributes. For example, "the Laplanders (a now-discarded European term for the indigenous Sámi peoples of northern Scandinavia and Russia) ... have the flat face and diminutive stature of the Samoyedes... In their dealings. ... they are described as mistrustful and given to cheating... The Samoyedes ... seem all of a heap; have short legs, small neck, a large head, flat nose and face (and are) extremely indifferent on all those subjects that excite the feelings of other people…. The Kamschatkans have the physical traits of the adjacent Polar tribes (and) despise labor, which they resume only from the necessities of the passing hour, and are cowardly in the extreme" (Morton *op cit.*: 51–52). And so it went. The 'Koriaks' ("dull of comprehension, obstinate and revengeful"), the 'Tchukches' ("barbarous and cruel, and repugnant to every form of civilization").

Returning to the North American Arctic, Morton found people "composed of the Eskimaux and Greenlanders, who are both generally included in the former name, an Algonkin word signifying "eaters of raw flesh," but their own national designation is Keralit (present-day Kalaallit). They are the sole inhabitants of the shores of all the seas, bays, inlets and islands of America, north of the 60th degree of north latitude, from the eastern coast of Greenland in longitude 21°, to the straits of Behring in longitude 127° west" (Morton *op cit.*: 52) (Fig. 2.1).

Morton's knowledge of Greenland seems to have derived principally from two sources, both of which he liberally quotes from. The Moravian church sent David Crantz to Greenland for a year in 1759. His account, published in German in 1765, was only translated into English in 1820, soon after a sort of craze over the subject of Greenland had begun in England (Crantz, 1820). The other was Wilhelm Graah's account of his 1829 expedition on behalf of the Danish crown 'in Search of the Lost Colonies' in Greenland, published in Danish in 1832 and appearing in English five years later (Graah, 1837).

> On the icy shores of the great island of Greenland, are seen the easternmost tribes of this singular race. Their features do not materially differ from those already described, but their complexion is decidedly darker, varying from brown to olive, while at Oppernivick (Upernavik) they are as dark as mulattoes. It is needless to add that many are much lighter, and others quite fair (Morton op cit.: 54).

Morton proceeds on this basis to his estimation of the Greenlanders place on "the moral scale" where they ranked "extremely low."

> They are crafty, sensual, ungrateful, obstinate and unfeeling, and much of their affection for their children may be traced to purely selfish motives. They devour the most disgusting aliments uncooked and uncleaned, and seem to have no ideas beyond providing for the present moment. With respect to the moral and intellectual character of this widely distributed family, little need be added to what has already been said. Their mental faculties, from infancy to old age, present a continued childhood: they reach a certain limit and expand no farther… They possess simplicity without silliness, and good sense without the art of reasoning. They are fickle and facetious, and their connubial infidelity is a proverb among voyagers. In gluttony, selfishness and ingratitude, they are perhaps unequalled by any other nation of people; and they are habitually unfeeling without designing to be cruel… Finally, though grossly puerile in their superstitions, they have, no combination of sentiments that deserves the name of religion (Morton op cit.: 54–55).

Morton then proceeds to his main point with regard to Greenland and Greenlanders. There were actually *two* Greenlandic peoples, the low gluttonous, ungrateful savages of the western coast and then the remarkably clear and pure of skin (and hence moral character) east coast Greenlanders encountered by Graah.

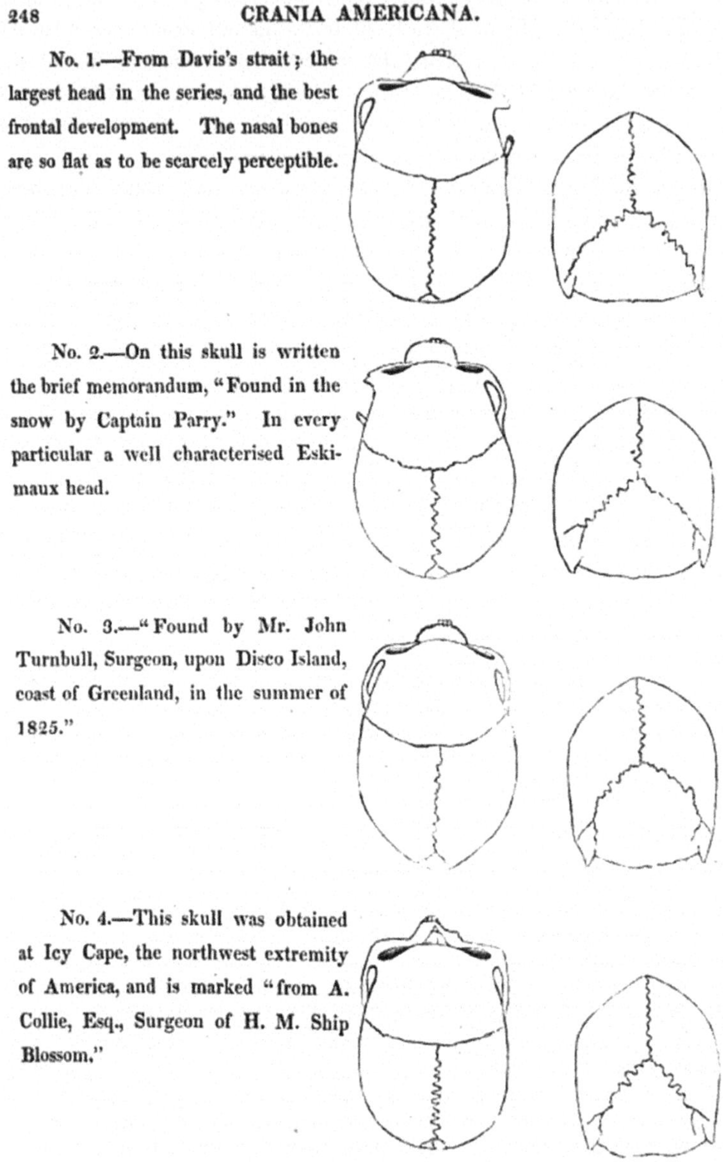

248 CRANIA AMERICANA.

No. 1.—From Davis's strait; the largest head in the series, and the best frontal development. The nasal bones are so flat as to be scarcely perceptible.

No. 2.—On this skull is written the brief memorandum, "Found in the snow by Captain Parry." In every particular a well characterised Eskimaux head.

No. 3.—"Found by Mr. John Turnbull, Surgeon, upon Disco Island, coast of Greenland, in the summer of 1825."

No. 4.—This skull was obtained at Icy Cape, the northwest extremity of America, and is marked "from A. Collie, Esq., Surgeon of H. M. Ship Blossom."

Fig. 2.1 Plate from Samuel Morton's *Crania Americana*. "Since writing the chapter on the Polar Family …. I have been favored by George Combe, Esq., with the use of four genuine Eskimaux skulls… The extreme elongation of the upper jaw contracts the facial angle to a mean of seventy-three degrees, while the mean of three heads of the four, gives an internal capacity of eighty-seven cubic inches, a near approach to the Caucasian average." (from: Morton, 1839, p. 247, 'The Mongol-Americans: Eskimaux')

> Most readers are aware that colonies of Scandinavians and Icelanders peopled Greenland in the middle ages. Since the fourteenth century, however, nothing has been heard of them, and they were supposed to have been blocked up and destroyed by the accumulating ice, whence the name of *Lost Greenland*. In 1829 the Danish government sent Captain Graah to explore these icy solitudes, and to ascertain at least the locality of the lost colony. This enterprising voyager discovered a community of which he gives the following account: "They have little analogy with the Eskimaux, and resemble, on the contrary, the Scandinavians of Europe. They have neither the flat heads, short broad persons nor flabby features of the Eskimaux; but are for the most part above the middle stature, having the European form of head and expression of countenance. Their persons are rather meagre, but nervous and finely formed, without any appearance of weakness, and they are more active and robust than the inhabitants of the western coast. The color of the skin of the women and children is quite clear and pure as that of Europeans, and they have often brown hair, which is never seen in the other inhabitants of Greenland." The moral character of these people is said to be characterised by great honesty, simplicity and truth: yet they are pagans, have their sorcerers like the Eskimaux, and speak probably a dialect of their language, for Captain Graah could not understand it (Morton op cit.: 55).

Armed with these "facts," Morton reaches his inescapable deduction. It was in fact these "finely formed .. active and robust" pure-skinned Greenlanders that had clung to survival on that icy shore from the time of Eírik raude. Further, any moral decay or descent into sorcery was solely due to the degenerative influences of the invading eaters of raw flesh from the high north. In less than a century, the inductive reasoning of Franklin had mutated to the facile deductions of Morton sprung from a ringing defense of the European colonial ideal.

Appropriately, Morton reached the only conclusion possible to him to explain the appearance of robust humans in Greenland. "It will be readily surmised from the preceding facts, that these people constitute the real remains of the Scandinavia colony which, to Europeans, have been lost for ages; and their long intercourse with the Greenland tribes has led them to adopt the superstitions of that people and more or less their language and mode of life" (Morton op cit.: 55–56).

Thus did Samuel Morton breathe life into the myth of the Lost Colony of Greenland and influence the next Philadelphian to link his name with the great island. And for Elisha Kent Kane, Greenland and its imagined gateway to the North Pole would become a path not to the top of the world but to a global fame the likes of which no American before him, even Benjamin Franklin, had achieved (Fig. 2.2).

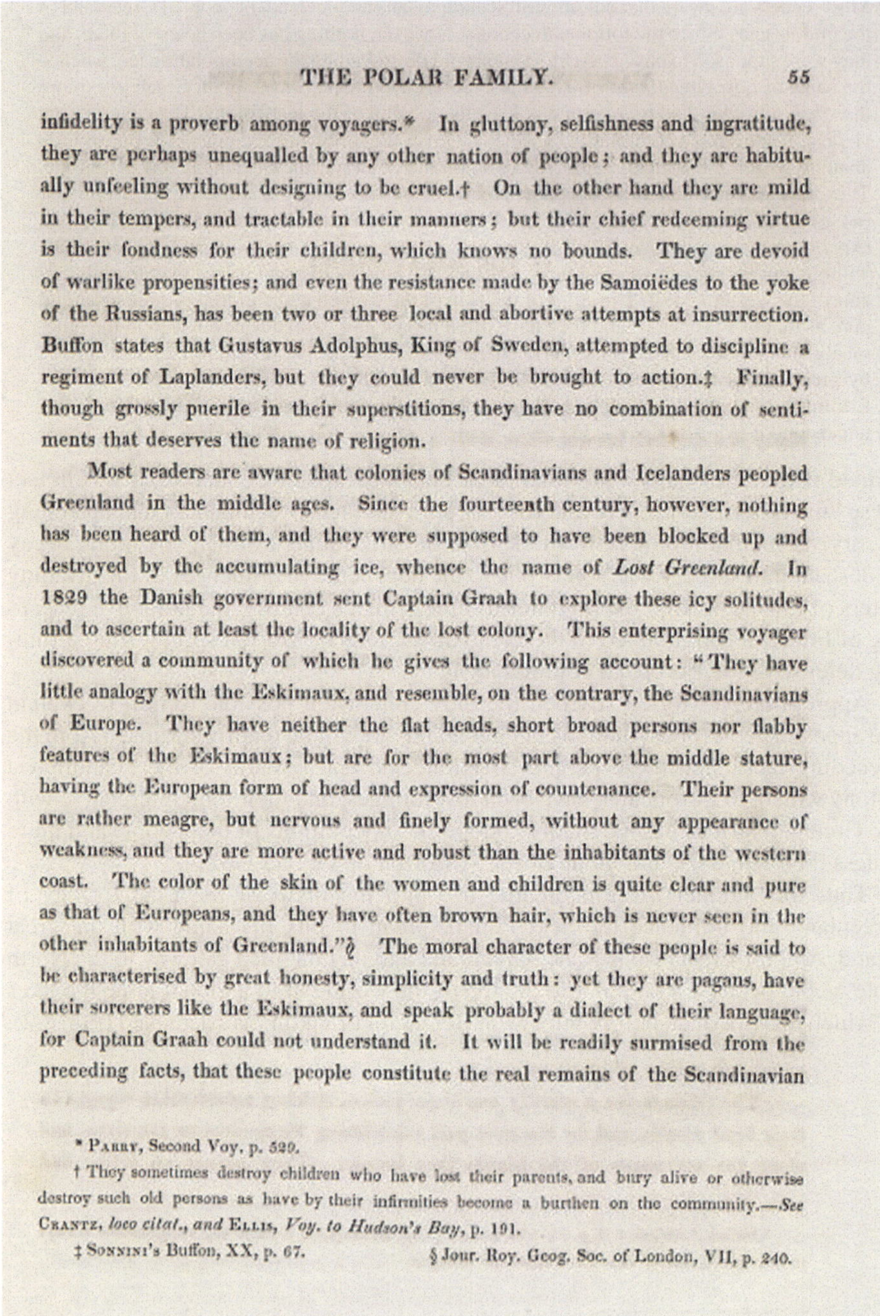
THE POLAR FAMILY. 55

infidelity is a proverb among voyagers.* In gluttony, selfishness and ingratitude, they are perhaps unequalled by any other nation of people; and they are habitually unfeeling without designing to be cruel.† On the other hand they are mild in their tempers, and tractable in their manners; but their chief redeeming virtue is their fondness for their children, which knows no bounds. They are devoid of warlike propensities; and even the resistance made by the Samoiëdes to the yoke of the Russians, has been two or three local and abortive attempts at insurrection. Buffon states that Gustavus Adolphus, King of Sweden, attempted to discipline a regiment of Laplanders, but they could never be brought to action.‡ Finally, though grossly puerile in their superstitions, they have no combination of sentiments that deserves the name of religion.

Most readers are aware that colonies of Scandinavians and Icelanders peopled Greenland in the middle ages. Since the fourteenth century, however, nothing has been heard of them, and they were supposed to have been blocked up and destroyed by the accumulating ice, whence the name of *Lost Greenland.* In 1829 the Danish government sent Captain Graah to explore these icy solitudes, and to ascertain at least the locality of the lost colony. This enterprising voyager discovered a community of which he gives the following account: "They have little analogy with the Eskimaux, and resemble, on the contrary, the Scandinavians of Europe. They have neither the flat heads, short broad persons nor flabby features of the Eskimaux; but are for the most part above the middle stature, having the European form of head and expression of countenance. Their persons are rather meagre, but nervous and finely formed, without any appearance of weakness, and they are more active and robust than the inhabitants of the western coast. The color of the skin of the women and children is quite clear and pure as that of Europeans, and they have often brown hair, which is never seen in the other inhabitants of Greenland."§ The moral character of these people is said to be characterised by great honesty, simplicity and truth: yet they are pagans, have their sorcerers like the Eskimaux, and speak probably a dialect of their language, for Captain Graah could not understand it. It will be readily surmised from the preceding facts, that these people constitute the real remains of the Scandinavian

* Parry, Second Voy. p. 529.

† They sometimes destroy children who have lost their parents, and bury alive or otherwise destroy such old persons as have by their infirmities become a burthen on the community.—*See* Crantz, *loco citat., and* Ellis, *Voy. to Hudson's Bay*, p. 191.

‡ Sonnini's Buffon, XX, p. 67.

§ Jour. Roy. Geog. Soc. of London, VII, p. 240.

Fig. 2.2 Page from Samuel Morton's Crania Americana, where the absence of Europeans from their Greenland colony left it with the name 'Lost Greenland'

References

Baker, P., & Glasser, S. (2022). *The divider: Trump in the white house, 2017–2021*. Doubleday. See also: Chitirala, Isha, and Finn Ryan. 2025. "How former donor, Penn graduate Ronald Lauder could be responsible for Trump's push to buy Greenland." Philadelphia: The Daily Pennsylvanian, January 27, 2025. Online at: https://www.thedp.com/article/2025/01/penn-trump-lauder-greenland-purchase

Benson, A. B. (1937). *Peter Kalm's travels in North America. The English Version of 1770* (2 vols., N.Y., 1937); and Esther Louise Larsen, trans. 1945. "Pehr Kalm's Report on the Characteristics and Uses of the American Walnut Tree Which is Called Hickory," *Agricultural History*, XIX (1945), 55–64. Online at: https://founders.archives.gov/documents/Franklin/01-04-02-0011

Crantz, D. (1820). *The history of Greenland: Including an account of the mission carried on by the united brethren in that country* (Vol. 2). Longman, Rees, Orme, and Brown.

Franklin, B.. 1739 Poor Richard, 1739. An Almanack for the year of Christ 1739, By Richard Saunders, Philom. Philadelphia:. Printed and sold by B. Franklin, at the New Printing-Office near the Market. (Yale University Library). Online at: https://founders.archives.gov/documents/Franklin/01-02-02-0046.

Graah, W. A. (1837). *Narrative of an expedition to the East Coast of Greenland, sent by order of the king of Denmark, in search of the lost colonies*. John W. Parker.

Jefferson, T. (1832). *Notes on the state of Virginia*. Lilly and Wait. Also: Jefferson, Thomas. 1785 (1975). *The Portable Thomas Jefferson*, edited by Merrill D. Peterson. New York: Penguin: 142–146. For a (mercifully) brief recap of the Solutrean hypothesis and its capture by extra-academic elements, see White, Andy. 2015. https://www.andywhiteanthropology.com/blog/does-the-fan-base-of-the-solutrean-hypothesis-change-if-upper-paleolithic-europeans-werent-white (accessed 18 February 2025)

Jordan, J. W. (1896). Moravian immigration to Pennsylvania, 1734–1767. *Transactions of the Moravian Historical Society, 5*(2), 51–90. Online at: https://www.jstor.org/stable/41179779

Kennedy, R. (2003). *Mr. Jefferson's lost cause: Land, farmers, slavery, and the Louisiana purchase* (p. 26). Oxford University Press.

Miller, J. L. 2017. "Our Own Flesh and Blood?: Delaware Indians and Moravians in the Eighteenth-Century Ohio Country.the Eighteenth-Century Ohio Country." Graduate Theses, Dissertations, and Problem Reports. 8183. https://researchrepository.wvu.edu/etd/8183.

Moravian Archives. (2011). Matthaeus Stach brings Greenlanders to Bethlehem. *This Month in Moravian History*. Bethlehem, PA: *Moravian Archives* 62 (March 2011). Online at: https://www.moravianchurcharchives.org/thismonth/11_03%20Stach.pdf. See also: Fischer, Taylor. 2023. "Paul Peucker, The Moravian Church in North America." Online at: https://benfranklinsworld.com/episode-356-paul-peucker-the-moravian-church-in-north-america/. The Moravians would continue as an active spiritual community in Greenland until 1900, which was something of a miracle in itself. The founding of a Moravian mission in Greenland was anything but a well-considered affair, not least on account of the missionary's complete ignorance of Greenland, how to reach it, or how to support themselves once there. See, for example, the remarkably blunt: Brown, William, and Adam Clarke. 1816. *The History of Missions: Or, Of the Propagation of Christianity Among the Heathen, Since the Reformation*, Volume 1 (pp. 290–370). Philadelphia: B. Coles. Online at: https://books.google.com/books?id=Vn8XAAAAYAAJ&pg=PA290#v=onepage&q&f=false.

Morton, S. (1839). *Crania Americana: A comparative view of the skulls of various aboriginal nations*. J. Dobson.

Rix, R. W. (2023). *The vanished settlers of Greenland: In search of a legend and its legacy*. Cambridge University Press.

Solis-Cohen, B. (1952). Philadelphia's expedition to Labrador. *Pennsylvania History, 19*(2), 148–162.

Solis-Cohen, B. (1943). "Philadelphia's Expeditions to Labrador." Pennsylvania History: A Journal of Mid-Atlantic Studies 19 (2):148–162.

Solis-Cohen, B. (1957). Benjamin Franklin defends Northwest passage navigation. *The Princeton University Library Chronicle, 19*(1), 15–33. autumn 1957. Online at: https://www.jstor.org/stable/26403271

Chapter 3
Elisha Kent Kane, the Open Polar Sea North of Greenland, and the American Route to the North Pole

Abstract

While we were standing upon deck, waiting for the boat to be manned which was to take us to the shore, something like a large Newfoundland dog was seen moving rapidly through the water. As it approached, we could see a horn-like prolongation bulging from its chest, and every now and then a queer movement, as of two flapping wings, which, acting alternately on either side, seemed to urge it through the water. Almost immediately it was alongside of us, and then we realized what was the much talked of kayack of the Greenlanders.

—Elisha Kent Kane, *The United States Grinnell Expedition in search of Sir John Franklin* (1857: 37)

Keywords Greenland · Greenlandic Inuit · Philadelphia · Grinnell expeditions · Disko · Kennedy Channel · Smith Sound · American Route · Etah · Annoatok

Samuel Morton's dim view of Arctic peoples prevailed at the very moment explorers from Philadelphia such as Edwin De Haven, Elisha Kent Kane, and Isaac Israel Hayes, inaugurated the nineteenth century American exploration of the Arctic. Morton's collection of skulls from around the globe eventually reached nearly 900 such objects, and after his death in 1851 the collection went to the Academy of Natural Sciences in Philadelphia. A century later, the collection became part of the University of Pennsylvania Museum of Archaeology and Anthropology.

Elisha Kent Kane himself graduated from University of Pennsylvania Medical College in 1842, at the height of Morton's influence on American intellectual life. Kane does not write of collecting the skulls of Greenlanders for Morton's research during the two Grinnell expedition launched ostensibly to search for the missing Sir John Franklin, but it would not be surprising. He had no qualms over otherwise disturbing Greenlandic graves. Going ashore to establish a cairn at Littleton Island during the second Grinnell expedition, Kane writes:

> These Esquimaux have no mother earth to receive their dead, but they seat them as in the attitude of repose, the knees drawn close to the body, and enclose them in a sack of skins. The implements of the living man are then grouped around him; they recovered with a rude dome of stones, and a cairn is piled above. This simple cenotaph will remain intact for generation after generation. The Esquimaux never disturb a grave.

P. J. Capelotti, *Lost Greenland*, Springer Polar Sciences,
https://doi.org/10.1007/978-3-032-27841-8_3

> From one of the graves I took several perforated and rudely-fashioned pieces of walrus ivory, evidently part of sledge and lance gear… We found … a child's toy spear, which, though elaborately tipped with ivory, had its wooden handle pieced out of four separate bits, all carefully patched and bound with skin (Kane, 1857: 31).

Perhaps the greatest irony in this particular cultural desecration might be that Kane's own gravesite was eventually vandalized. Dead of a weakened heart and rheumatic fever at just 37, Kane was given a national mourning that included a cross-country railroad cortège the likes of which was bested only two times in American history, for the funerals of Abraham Lincoln and Robert F. Kennedy. He was interred in a sealed metal casket in his family crypt in Philadelphia that was fit for an Egyptian pharaoh. Kane's crypt was last entered about twenty years ago—to reveal that the glass windows above his face and hands had been smashed, presumably by vandals from decades earlier who wreaked havoc on the tomb before it was permanently sealed (Capelotti, 2016).

Whoever broke into the crypt and scattered Kane's bones probably had little idea of who he was or why he had been buried in such an elaborate tomb, any more than Kane himself possessed knowledge of the Greenlandic peoples whose graves he disturbed. "We found to our surprise that we were not the first human beings who had sought a shelter in this desolate spot. A few ruined walls here and there showed that it had once been the seat of a rude settlement; and in the little knoll which we cleared away to cover in our storehouse of valuables, we found the mortal remains of their former inhabitants" (Kane *op cit.*) (Fig. 3.1).

Kane was born in 1820, just two years after the start of a cultural craze over Greenland that particularly infected England. An 1815 volcanic explosion in the then-Dutch East Indies had cast a pall over the global environment, leading to lower temperatures, crop failures, and food riots in the so-called Year Without a Summer of 1816. Observations of possible ice-free zones around Greenland by William Scoresby, Jr., in 1817, were seen as a potential way out of an apparently destabilized global climate. These in turn were seized upon by the post-Napoleonic wars British Admiralty, which in 1818 took up Arctic expeditions in search of a Northwest Passage to the markets of Asia, voyages that had been left aside since the Arctic voyage of Constantine Phipps in 1773.

In the summer of 1818, an expedition led by John Buchan ranged around Spitsbergen (Svalbard), while another of John Ross explored the west and northwest coast of Greenland, where Ross famously encountered High Arctic Greenlanders. This meeting of a British naval vessel with High Arctic Greenlanders also revived the issue of the 'Lost Colony,' which had become the subject of intense discussion. The Admiralty's John Barrow published an extended history of Arctic exploration in 1818, in which Greenland is presented as "a geography that falls within the British sphere of interest" (Rix *op cit.*). A year earlier, Barrow's article anonymously published in the *Quarterly Review* explicitly linked British Arctic exploration with the search for the Lost Colony.

Fig. 3.1 Image of Kane's tomb at Laurel Hill Cemetery, Philadelphia (from Elder 1857: 247)

> It is generally admitted that, for the last four hundred years, an extensive portion of the eastern coast of Old Greenland has been shut up by an impenetrable barrier of ice, and, with it, the ill-fated Norwegian or Danish colonies, which had been established there for more than an equal length of time preceding that unfortunate catastrophe, and who were thus cut off at once from all communication with the mother-country;—that various attempts have been made from time to time to approach this coast, with the view of ascertaining the fate of the unfortunate colonists, but in vain, the ice being everywhere impervious; and that, all hope being at length abandoned, that part of this extensive tract of land which faces the east took the appropriate name of *Lost Greenland* (Barrow, 1817: 199–223).

Barrow's agitation for exploration was matched by the popular culture of the day. Mary Shelley's *Frankenstein; or, The Modern Prometheus* appeared in 1818, as did Caspar David Friedrich's *Wanderer above the Sea of Fog*. Both presented men seemingly in control of their environment, and both seemingly about to tumble into catastrophe. Shelley's monster wins a race towards the North Pole with its doomed creator, while just five years after *Wanderer*, Friedrich painted *The Sea of Ice*, where humans are nowhere to be found in a jagged landscape of twisted pack ice, and it takes some effort to recognize the crushed remnants of ships destroyed by the lethal Arctic.

Critics of the day were unimpressed. Friedrich's brilliance remained unrecognized until the twentieth century, and an 1818 reviewer of Shelley's now iconic work remarked acidly that it was little more than "a tissue of horrible and disgusting absurdity..." (Anonymous, 1818a, b: 379–384). Barrow fared little better. In an essay for the *Literary Journal* of April 1818, an anonymous writer took direct aim at the second Secretary of the Admiralty and declared: THERE IS NO "LOST GREENLAND."

> The notion of an European colony, anciently subsisting on the east coast of Greenland, and, for four centuries past, cut off from all communication with its parent country, and, perhaps, totally destroyed by accumulation of ice—this notion ... lately revived in the *Quarterly Review*, where it gives so striking and melancholy an interest to the speculation on the present possibility of visiting the site of that colony, and navigating what are assumed to be the "Polar Seas:"--this notion, which has been made, we say, to confer a romantic interest in that speculation, by holding out to us the prospect that our seamen may find some remains of that colony — some remains even of its population — appears wholly to vanish, upon being, with a little care, investigated (Anonymous, 1818a, b: 40–42).

Elisha Kent Kane grew up surrounded by such debates. When he was only five, his father, John Kintzing Kane, a Yale-educated lawyer who would rise to become Pennsylvania's attorney general and a Federal judge, was elected to the American Philosophical Society (APS) and late in life served as its president, a position Thomas Jefferson had held for 17 years, Benjamin Franklin for twenty-one. (Samuel Morton was elected a member three years after Kane's father.)

Kane's family affluence trace to his grandfather, a charter director of the Philadelphia Bank, and his father, associations that would be used by Kane in gathering scientific support for his Arctic expeditions. Kane's father built two homes in North Philadelphia, "Rensselaer" on Green Lane in 1848 and "Fern Rock" near Chelten Avenue and Old York Road in 1853. His son, the future explorer, would soon attach these names to a harbor and an island, respectively, discovered during his explorations of Greenland, and the Fern Rock neighborhood of Philadelphia is named for the Kane family home (see, for example, Corner, 1972: 3–31; McGoogan, 2008: 15–27; Elder, 1858: 20–43).

John Kane wanted his eldest son to be a surgeon, but Elisha Kane desired a life in the field and began studies at the University of Virginia in geology and civil engineering. Ill health forced his return to Philadelphia—in the same year Morton published *Crania America*—and desk-bound medical studies at the University of Pennsylvania, where he graduated in March 1842 as the foremost student in his class. His father secured him a commission as an assistant surgeon in the U.S. Navy, attached to the Philadelphia Navy Yard. But Kane used the position to embark on a brief but intense life of adventure in both the tropics and, most famously, in the Arctic.

We might trace the moment that all Americans not named Franklin, Jefferson, or Morton began to care about the Arctic to the morning of Tuesday, June 3, 1845, during the middle of Kane's world travels as an assistant naval surgeon. That was when a brief notice appeared on the front pages of American newspapers announcing the British expedition of Sir John Franklin and the ships *Erebus* and *Terror*. By the end of that summer, over 120 officers and men under Franklin had sailed into the tangle of islands north of Canada and vanished. By 1849, just as Kane began work surveying the southeast coast of America for the U.S. Coast Survey, it was clear that a catastrophe had occurred as the British Admiralty sent eight more ships in search of the original two and then offered a huge reward for the rescue of Franklin and his men. Here was a heaven-sent opportunity for an ambitious young man in naval service (Fig. 3.2).

Franklin's wife made impassioned appeals for even more ships to be sent in search of her husband and his men, with petitions to the British government, to Russia, and then to the United States. Twice in 1849 she wrote long letters to President Zachary Taylor, missives that successively begged, bribed, shamed, and cajoled the country into action. The letters also had the effect of encouraging a New York shipping merchant named Henry Grinnell to offer the U.S. government two of his own vessels for the task. Congress thereupon agreed to staff the ships with officers and men of the U.S. Navy.

Command of the expedition was given to Edwin De Haven, then a thirty-four-year-old lieutenant who had served on Charles Wilkes' Great U.S. Exploring Expedition from 1838–1842, an expedition that had announced America as a nation in search of geographic, biological, and ethnological knowledge. A British surgeon serving in the Franklin search described De Haven as "as fine a specimen of a seaman, and a rough and ready officer, as I had ever seen" (Gleaves, 1928).

De Haven had entered naval service as a midshipman at the age of thirteen in 1829. He served on board a series of warships in the West Indies and Brazil before being assigned to the Wilkes expedition in 1839. It was during this expedition when De Haven was credited with saving several lives when the sloop-of-war U.S.S. *Peacock* grounded and wrecked at the mouth of the Columbia River. De Haven returned from the Wilkes expedition to regular naval duty, concluding with a tour at the U.S. Naval Observatory in Washington, D.C. It was in Washington in

Fig. 3.2 Edwin Jesse De Haven, commander of the First Grinnell Expedition (public domain via Wikimedia)

March of 1850 where De Haven received a letter from Henry Grinnell selecting him to command the first American search for the Franklin expedition.

De Haven responded to Grinnell immediately that he "accepted command of the Arctic expedition … unconditionally and without hesitation." In addition to a standard cohort of junior officers, De Haven requested a medical officer as well. This request would lead to the appointment of his fellow Philadelphian Elisha Kent Kane as surgeon on the expedition.

A steam tugboat towed the tiny convoy from New York Harbor on May 22, 1850 and two and half weeks later the two vessels arrived in Newfoundland. From there, the ships sailed towards Greenland, which they sighted on June 20. Here Kane began his observations in earnest, remarking on driftwood and ice floating into Davis Strait "evidently from the Spitzbergen Seas" (Kane, 1854: 30). They entered Disko Bay, since Edward Parry's visit in 1821 "an established resort for whalers and expedition ships" (Kane *op cit.*: 36) and found that a British expedition in search of Franklin under the command of Horatio Austin had just the previous day departed the port for the Canadian Arctic.

At Disko, Kane marveled at water so clear one could look down ten fathoms and see every feature of the bottom. He saw a kayak for the first time, describing it as "graceful as the nautilus" (Kane *op cit.*: 37) and so harmonious a construction that it was "hard to say where the boat ended or the man commenced" (Kane *op cit.*: 38) To the humans who had created these marvels, however, Kane echoed the cultural superiority of Samuel Morton. Where a more sensitive observer might note the remarkable ability of these small family groups to survive in the open where no others had managed, either before or since, Kane described his "disgust" at dwellings he saw as "mouldy, dank, and fetid."

> The one feeling which, I venture to say, pervaded us all, to the momentary exclusion of every thing else, was disgust. Offal was strewn around without regard to position; scabs of drying seal-meat were spread over the rocks; oil and blubber smeared every thing, from the dogs'coats to their masters´; animal refuse tainted all we saw; and we afterward found, while botanizing among the snow valleys, bones of the seal, walrus, and whale, buried in the mosses. But if filth characterized the open air, what was it in the habitations! One poor family had escaped to their summer tent, pitched upon an adjacent rock that overlooked the sea. Within a little area of six feet by eight, I counted a father, mother, grandfather, and four children, a tea-kettle, a rude box, two rifles, and a litter of puppies (Kane *op cit*.: 39–40).

From Disko, the Arctic cruise truly commenced but spent virtually the entire month of July caught in ice of Melville Bay. The expedition reached its furthest north in Baffin Bay, at 76° 25′, in mid-August, when open waters allowed them to enter into the Canadian Arctic where the two vessels separated. On the 19th, the Americans fell in briefly with another British search expedition, this one led by the Scottish captain William Penny. A few days later they met yet a third search expedition, the *Prince Albert* under Charles Forsyth.

In late August, maneuvering in the ice of Radstock Bay, De Haven learned that *Rescue* had discovered evidence of the Franklin expedition a few miles to the west on Beechey Island. Along with the British, the two vessels proceeded to the spot, which Kane described as more of a peninsula, with finds including newspaper fragments from 1844 and bits of written documents containing the names of some of Franklin's officers. Kane thought Franklin must have moved on, as a party as large as his would certainly have left behind more evidence of their encampment had they remained in an area alive with birds, polar bear, Arctic fox, and whales (Fig. 3.3).

Just as the British and American captains conferred on a division of labor in their continued searches, word came that three graves had been found at Beechey. Rushing to the scene, the officers found the remains of three of Franklin's men, sailors who had been buried four and half year earlier. As Kane recorded, additional

Fig. 3.3 Drawing of the graves from the Franklin expedition on Beechey Island, from a sketch by Elisha Kent Kane (from Kane, 1854)

finds included "a large blackened space, covered with coal cinders, iron nails, spikes, hinges, rings, clearly the remains of the armorer's forge" (Kane *op cit*.: 164) Nearby were the remains of a carpenter's shop, a rubbish dump with empty tins of preserved meats, and the odd scraps of paper, sails, blankets, socks, and mittens.

It seemed clear to Kane that Franklin had made a systemic survey of the area before attempting to move further along the Northwest Passage. But where had he gone from Beechey? Kane's imagination on these possibilities would lead to both his second Arctic expedition and ultimately to his early death as he labored over a book about that expedition. His wildest speculation was that Franklin had gone north to penetrate a mythical open polar sea and subsequently found himself trapped by a wall of ice that blocked his retreat south. If one could only sail north, while always keeping the coast of Greenland on the starboard to be used as a shield from the crush of ice circulating around the North Pole, one could both locate Sir John while sailing or sledging to the Pole itself.

Captain De Haven was infinitely more clear-eyed and clear-headed. With the onset of cold weather in mid-September, he realized that any further passage west was out of the question. He tried to lead his exploring convoy back eastwards toward Baffin Bay but was stopped by the ice. The ships drifted north where they sighted land later named Grinnell Peninsula after the expedition sponsor and then drifted south where the movement of pack ice shifted them eastwards a few miles each day throughout the winter and spring of 1851. The officers and men increasingly suffered from scurvy, as Kane lamented the lack of fresh food to keep them healthy.

When the ships finally reached Baffin Bay and broke out of the ice in early June, De Haven found refuge in Greenland so that his men could rest and recover. Two weeks later, once again unable to maneuver in thick ice, De Haven ordered the ships home.

Back in Pennsylvania, De Haven gave the job of writing up the expedition narrative to Kane. Kane's report on this first significant American foray into the Arctic would make him the expedition's most famous veteran and, in the eyes of many of his fellow Americans, the most famous explorer in the world. Kane had been elected a member of the APS in May of 1851 while he was still in the Arctic. He made his first formal address to the society a year later, on Arctic flora, followed by another lecture four months later, on the drift of ice in Baffin Bay. At the age of 32, Kane had entered the American scientific elite. His account of his first expedition to the north framed the Arctic as a region where humans could engage in a transcendent struggle against implacable, otherworldly elements, and this Romantic portrait would influence both scientific and popular thought about the Polar Regions for nearly a century.

Kane could barely contain his enthusiasm to return. His second expedition in the brig *Advance* gained the full support from another fellow Philadelphian and Superintendent of the Coast Survey Alexander Dallas Bache, as well as the Secretary of the Navy, John P. Kennedy. The APS raised funds for the expedition and the Academy of Natural Sciences in Philadelphia planned the biological research program for the cruise. Scientific instruments were recommended by the navy's influential oceanographer Matthew Fontaine Maury.

Kane recruited fellow Pennsylvanians to join him, including Amos Bonsall, a farmer from Delaware County, as well as a University of Pennsylvania medical student named Isaac Israel Hayes of West Chester, who became the expedition's surgeon and natural historian. Kane's plan, to use the west coast of Greenland as a blocking arm to shield his ship as it traveled north, relied also on the cooperation and local knowledge of Greenlanders. Still, Kane would later see no contradiction in robbing the graves of these same people to fill museum collections in Philadelphia.

During this voyage, Kane charted the coasts of Smith South and what would be posthumously named Kane Basin and ventured farther north than any previous explorer, into a channel he named for Secretary Kennedy. Kane's book describing this expedition, *Arctic Explorations*, has been described as "the best written and most readable book on [the] subject in the English language" and "a masterpiece of travel literature" (Corner *op cit.*: 258, 269). His use of a small, maneuverable vessel in the Arctic, an approach advocated by De Haven, was taken up by subsequent explorers like Roald Amundsen and Donald MacMillan. Nothing, however, so influenced the course of America's fixation with Greenland over the next 75 years as Kane's imagination of open, ice-free water north of the great island.

Kane was also aware that his supplies were barely adequate for one wintering in the Arctic, much less the two that the expedition eventually became. "Our stock of fresh meats was too small, and we had no preserved vegetables" (Kane *op cit.*: 16). He refused, however, to hold himself responsible, writing that "his personal means were limited" (*ibid.*) and he was reluctant to approach his sponsors for more support than they had already provided. His crew—along with Kane himself—would pay a heavy price in starvation, scurvy, and, for three of them, with their lives.

The new expedition sailed from New York on the *Advance* on 30 May 1853 and after a stop in Newfoundland made directly for the southwest coast of Greenland. There Kane embarked a nineteen-year-old Greenlandic hunter and dog driver named Hans Hendrik, who would become a fixture on American Arctic expeditions for the next thirty years. His Greenlandic name was Suersaq, and he authored the first native Greenlandic account of being on an expedition with whites from the south. Kane may have been revolted by Greenlandic foodways, but fortunately for him and the other American survivors of the doomed expedition, Hendrick was superbly adapted to survival on his native coast.

"The year after my father's death I departed, this first time joining Arctic travellers, an American sailing vessel," he remembered in 1878.

> "I heard that they were looking for a native companion, and that his parents should have payment during his absence. Nobody being willing, I, Hans Hendrik, finally took a liking to join them, and I said I would go...
>
> At Upernivik (Upernavik) we engaged as our companion a Kavdlunak [European], Carl Petersen, whom the natives called Naparsisortak [the new cooper] a very good Kavdlunak he was, and he took care of me. When we came to Upernivik, I was invited to eat in the great house, as he lived in the merchant's house. He was married to the midwife, and she and their children all treated me kindly. Leaving Upernivik we went very far to the north, and we landed at a desert north of where people lived it was not far from an abandoned winter station. When we had landed we fell in with the carcass of a white whale. Although pretty fresh, the birds had eaten the upper part of the skin, but the lower part was still

> untouched, and Naparsisortak and I got a store of matak [eatable skin]. Here we stayed, I think, for three days, and boiled the oil out of it. Then again we started, but, meeting with heavy drift-ice, made the land, and came to a native house, the northernmost of the many wintering stations thereabouts: its name was Anoritok (Annoatok) (Hendrick, 1878: 22–23).

From Hendrik's village of Fiskernæs, Kane directed his ship north, stopping to assemble dozens of cairns marking his progress while turning over native Greenlandic graves as he went. By early August, hundreds of miles further north, he deposited another cache of supplies along with a boat at Littleton Island (which would become a regular stopping point for future Americans), making sure, as we have noted, to rob artifacts from an Inughuit gravesite before he left (Fig. 3.4).

Kane was now at the entrance to Smith Sound, and to his mind the true start of the expedition. He had staked his career and the lives of his crew on a theory that somewhere to the north lay the opening to an open polar sea shielded by a west coast of Greenland that stretched to the North Pole. But as the *Advance* rounded the point of land near Etah and entered Smith Sound proper, Kane emerged from breakfast on Sunday, 7 August, ascended to the crow's nest, "and saw to my sorrow the ominous blink of ice ahead" (Kane, 1857: 30) This sight was compounded by a steady north wind driving that ice directly into the path of Kane's ship.

Fig. 3.4 Meeting the Esquimaux, from Kane's Arctic Explorations: The Second Grinnell Expedition, p. 202 (1857)

At the northern edge of the bay, the ship eventually pushed through Smith Sound and into a basin that was eventually named for Kane himself. A few miles further along the coast and the *Advance* entered into winter quarters at an indentation in the shore that Kane named Van Rensselaer Harbor after the maiden name of his father's mother. A cairn was erected that still marked the spot more than a century later. A storehouse was erected on an islet named Butler Island, while another islet Kane named Fern Rock after his Philadelphia birthplace. His Greenlandic Fern Rock now became the site of the expedition's observatory.

Despite a routine of hunting, cache-laying, and scientific observations, the condition of the men and the dogs worsened throughout the winter. By March of 1854 most of the dogs were dead and from an advance party of seven men only three returned to the ship. The other four had to be rescued off the ice and one of these, a childhood friend of Kane's named Jefferson Baker, died of infection on 8 April after Kane was forced to amputate one of his feet above the ankle. A second member of the party, the expedition's cook, Pierre Schubert, died in similar fashion a few weeks later. The remains of the two men were placed in coffins inside the observatory hut on Fern Rock, but the rocky islet admitted no possibility of a formal burial.

Kane himself set out on 26 April in an attempt to reach the North American side of Smith Sound but turned back after a week when advance depot were found to have been raided and destroyed by polar bear. Kane did, however, obtain his first clear views of the Humboldt Glacier, and his observations and sketches of this massive formation rising 300 feet above the surface of the frozen sea would condition glacier studies and the understanding of the interior of Greenland and the possibilities for a global Ice Age for the next hundred years. Kane paid the price for these observations in a worsening of his preexisting endocarditis that resulted in his return to the *Advance* in a state of delirium (Fig. 3.5).

A renewed hunt for seal meat in May led to a reduction in scurvy and led Kane to imagine that Franklin must yet be alive as he would have done the same. But Kane was forgetting the vast size of Franklin's expedition, which made it impossible solely to live of the land. Kane's own medical roster, provided by the twenty-three-year-old expedition surgeon Dr. Hayes as he returned from a survey across Kane Basin towards Cape Sabine in early June, showed that Kane was barely keeping the remainder of his small crew alive.

Kane was determined to learn what lay beyond the nearly seventy-mile-wide Humboldt Glacier and whether any waterway led north out of the basin in which *Advance* was frozen. As he himself was unfit for further fieldwork, he sent his six remaining healthy men on a two-team relay. Five men led by William Morton advanced to the glacier, there to be joined by Hans Hendrik. From there, Morton and Hendrik would form a two-person reconnaissance team in an attempt to see what lay to the north. It was to prove the most fateful foray of the entire expedition.

The two men proceeded beyond the Humboldt Glacier, and then still another glacier front, until on June 21 they rounded a point and in heavy mists realized they were standing on exceedingly thin ice in the middle of a fast-moving channel. When the mist cleared briefly, the men saw clear water running rapidly northwards. They rounded the cape, calculated their position at more than 81° North,

Fig. 3.5 Kane's sketch of the 'Great Glacier of Humboldt,' from Kane's *Arctic Explorations: The Second Grinnell Expedition* (1857)

and witnessed nothing but open water even further to the north. The sight seemed to confirm everything that Kane believed about Arctic geography. When Morton and Hendrik returned to the *Advance* and were able to report their observations, it was all Kane could do to contain his excitement in the belief that they had discovered the entrance to a navigable Arctic ocean. Kane's thoughts were soon to attain an almost mystical significance as the Open Polar Sea north of Greenland (Fig. 3.6).

Kane named the channel at the northern end of Kane Basin after his patron and supporter in the navy department, John Pendleton Kennedy. In his account of the voyage, Kane ecstatically described how Morton and Hendrik stood at 81° 22' North latitude, a point Kane named and claimed for America as "Cape Constitution," where they heard "the novel music of dashing waves; and a surf, breaking in among the rocks" (Kane *op cit*.: 181). This sound of soughing waves, Kane believed, was caused by his open, ice-free ocean at the top of the world. If one could but reach these waters, they would allow a ship direct passage to the North Pole. If Sir John Franklin and his surviving crew still lived—or even the surviving Norsemen of the Lost Colony—Kane was convinced it was there, on the shores of that open sea—or on an easily accessible island within it.

Kane was 30 years old when he first witnessed the Arctic with De Haven. By the time he was buried on the bluff overlooking the Schuylkill River less than seven years later, his imagination of an 'open polar sea' had gained a permanent place as siren song of nineteenth century theories of Arctic geography. Kennedy Channel, Cape Constitution, Kane Basin, all thereafter became the 'American route' to the North Pole. His own surgeon Hayes, followed by countrymen Charles Francis Hall, Adolphus Greely, Robert Peary, Frederick Cook, all at one time or another would

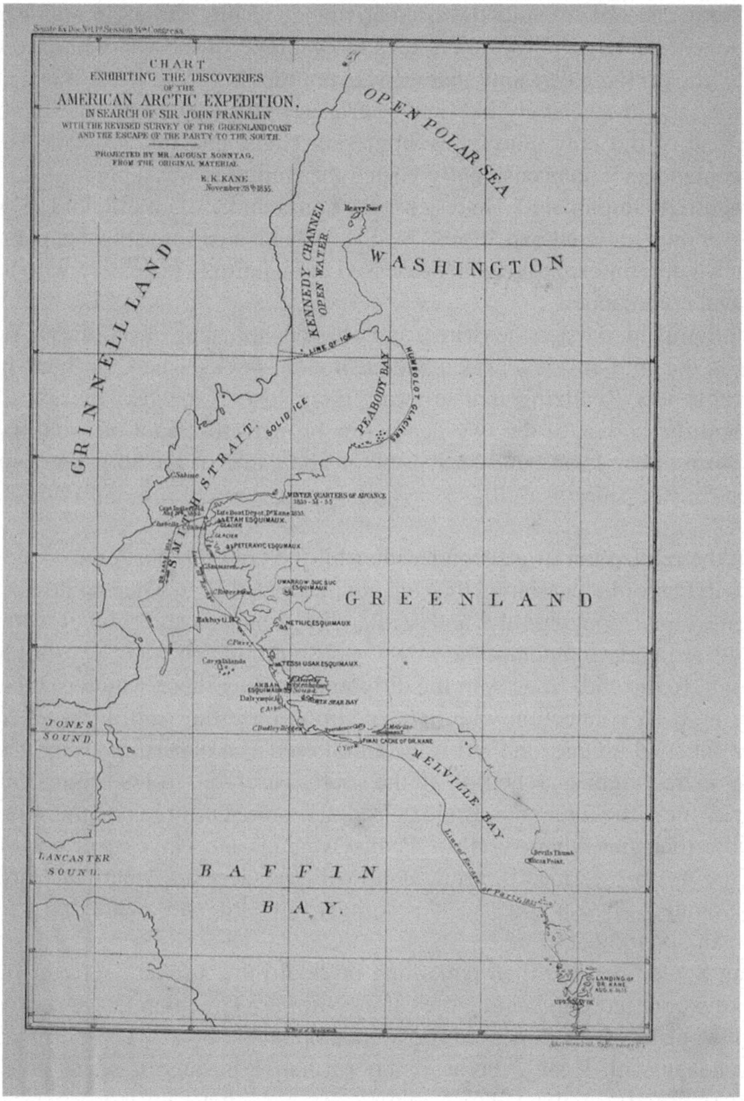

Fig. 3.6 "Chart exhibiting the discoveries of the American Arctic expedition in search of Sir John Franklin: with the revised survey of the Greenland Coast and the escape of the party to the south/ Projected by Mr. August Sonntag, from the original material; E.K. Kane." In the digital collection *UM Clark Library Maps*. https://quod.lib.umich.edu/c/clark1ic/x-003279950/39015091192925 (University of Michigan Library Digital Collections, Accessed January 14, 2026)

claim it—by right as Americans—as their own. The fact that Kane Basin and the Kennedy Channel—as Kane himself would soon demonstrate—were, in reality, little more than an ice-clogged trap for sailing ships barely made a dent in the calculus of these otherwise rational explorers.

The Arctic coast of Greenland wasted no time in testing Kane's dramatic hypothesis. Kane and his largely invalid crew, as the summer of 1854 wore on, were confronted by the growing certainty that ice was not moving from Kane Basin and that the *Advance* was increasingly likely to be frozen in for a second winter. Even with the expertise of the now thoroughly homesick Hans, and the assistance of more local Greenlanders who occasionally visited the ship, this was a prospect for which Kane was utterly unprepared. He was not the commander of an official U.S government vessel on a national expedition. Neither was he a private ship's captain in the form of a whaler, for example, who possessed more latitude in dealing with his crew than a naval commander.

Kane attempted to reach the British and a hoped-for cache of supplies at Beechey Island, but the mid-July excursion failed after 3 weeks when the men ran into impenetrable ice. Realizing that all was likely lost, Kane had the expedition's records and the bodies of the two dead crew brought to a spot on what had been named Observatory Island and buried there. The name of the ship was painted in tall letters on the side of a cliff, as a signal to anyone who might come looking for them.

When the realization of a second winter hit the crew in late August, more than half were in favor of abandoning the ship and the expedition. Kane could not decide if the men were disobeying rules and regulations as on a naval vessel or were in fact free agents at liberty to attempt their own way south once it was clear the ship was not moving. In the end, Kane split the difference, aiding those who wished to leave but severing their connection with the expedition once they did. It was a compromise that satisfied no one, and left Kane embittered as a departing group that eventually numbered eight marched off to the south. He wrote to his brother that if he should ever meet the three men Kane considered more cultured and therefore more responsible than the others—fellow Pennsylvanians Dr. Hayes and the Master's Mate of the *Advance*, Amos Bonsall, along with the German-American astronomer August Sonntag—he should "seek a solitary orchard and maul them for me" (McGoogan *op cit.*: 221).

During September, the men remaining on board the *Advance* took to stripping the ship of wood to use in heating the cabin. Northern Greenlanders appeared and a loose trade began which increased the food supplies on board. These irregular visits—that began with Kane's threat to shoot a native hostage if some stolen tools were not returned—evolved into a mutual aid pact that saw the Americans and Greenlanders working together on walrus hunts until the cold of October drove the wildlife away or into hibernation, and the northern Greenlanders back to their settlements.

In early December, Kane was stunned when the Inughuit returned to visit the ship with Bonsall and another crewmember who had left the ship in August. The remainder of the departure group were sheltering and struggling to survive the winter some 350 miles to the south. Kane sent supplies to them at once and they eventually all made their way back to the ship, debilitated and chastened at learning the truth of Kane's warnings against challenging the horrific ice conditions in the area. Again, Kane was forced to amputate appendages, this time several of Hayes' toes.

Sonntag could barely walk. The others suffered from contracted tendons and open sores that turned the *Advance* into a floating hospital.

Ten days later, a fire broke out that nearly consumed the ship. That crisis averted, the crew continued to suffer from scurvy and malnutrition throughout the winter. By the spring, of the eighteen survivors, only six were in any condition to work. On May 20, Kane buttoned up the ship and dragged the men to a small hut/shelter at Annoatok a few miles south.

With the help of Inughuit Etah, Kane began a shuttle of the survivors south. He also made repeated returns to the ship to use up the remainder of the expedition's flour in baking hundreds of pounds of bread to feed the retreating crew. All of this effort was not enough to save a third crew member, thirty-six-year-old carpenter Christian Ohlsen, who fell through the ice and died from the exertion of trying to regain the shore.

For their help in saving the remainder of the expedition, the Christianized Hans offered the northern Greenlanders—surviving for thousands of years without wood, without anything connected with the world beyond their ice—a sermon on the fear of otherworldly beings nailed to wooden crosses who would pursue them even unto death.

> It is a great pity that people there in the north have no idea of a Creator. Only by me were they informed about the Maker of heaven and earth and everything else, of all animals, and even of ourselves. Also, about His only begotten Son, who came in the flesh for the sake of sinful men, for the purpose of saving them, teaching them faith, and performing wonderful deeds amongst them, and afterwards was killed on a wooden cross, and arose from the dead on the third day, and will come down again to judge the living and the dead. On hearing this the Northlanders were rather frightened as to the destruction of the world in their life as well as in their death (Hendrick *op cit.*; 34).

Confronting hundreds of miles of densely packed ice in open boats and damaged sledges, Kane led his group south, at times walking, other times sailing or rowing. His natural history specimens and scientific notebooks were lost along the 1300-mile journey. The graves of the Greenlanders had been looted for nothing, except perhaps to fulfill the prophecies of Hans. The men eventually reached the broken pack of upper Baffin Bay and began to drift southwards with it. On 6 August 1855, their ordeal ended as they dragged their boats ashore at Upernavik.

Recuperating in the small Danish village, Kane learned that the Scotsman John Rae, after exploring the area around King William Island, had purchased Franklin relics from Inuit of northern Canada and effectively solved the mystery of the disappearance of the Franklin expedition. The shocking implication from Rae's report was that some of Franklin's men had been driven to cannibalize each other to survive. It was a fate that a third wintering might have forced upon Kane and his men.

A Danish vessel transported Kane and his survivors to Disko Bay, where they met an American search vessel that transported his surviving crew to New York. They arrived there on October 11, 1855. Kane had been forewarned by his brother that in his absence he had been made a national celebrity, and Kane dutifully played the part of modest returning hero, aware not to make any sweeping geographical claims or pronouncements.

Kane's brother also advised him to speak liberally and courteously to the press, which he did. He began work on the two-volume expedition account that would forever seal his reputation as America's first and foremost Arctic explorer. His subsequent account of Morton's discovery of Kennedy Channel and the presumed open sea beyond, would condition much of American Arctic exploration for the next half century, beginning with the Arctic explorations of his own problematic expedition surgeon, Isaac Israel Hayes.

References

Anonymous. (1818a). Frankenstein, or the modern Prometheus. *Quarterly Review, 18*, 38. (January 1818). Retrieved October 21, 2025, from https://babel.hathitrust.org/cgi/pt?id=hvd.32044092624956&seq=9

Anonymous. (1818b). There is no "lost Greenland.". *The Literary Journal, 3*. (April 12, 1818). Retrieved October 21, 2025, from https://archive.org/details/sim_literary-journal-and-general-miscellany-of-science-art_1818-04-12_1_3/page/40/mode/2up?q=lost+greenland

Barrow, J. (1817). Article XI. Narrative of a voyage to Hudson's bay, in his majesty's ship Rosamond, containing some account of the North-Eastern Coast of America, and of the tribes inhabiting that remote region. *Quarterly Review, 18.35*. (1817: Oct.). Retrieved October 21, 2025, from https://babel.hathitrust.org/cgi/pt?id=hvd.32044092624956&seq=210

Capelotti, P. J. (2016). The state of Arctic History: the preeminent polar explorers of Pennsylvania. From: *Our Arctic Nation*, published by the U.S. Department of State. Retrieved October 21, 2025, from https://medium.com/our-arctic-nation/week-29-pennsylvania-a7d0fe44e217

Corner, G. W. (1972). *Doctor Kane of the Arctic seas*. Temple University Press. McGoogan, Ken. 2008. *Race to the Polar Sea*. Berkeley, CA: Counterpoint; Elder, William. 1858. Biography of Elisha Kent Kane. Philadelphia: Childs & Peterson; Sawin, Mark M. 2008. Raising Kane: *Elisha Kent Kane and the Culture of Fame in Antebellum America*. Philadelphia: American Philosophical Society.

Elder, W. (1857). Biography of Elisha Kent Kane. Philadelphia: Childs & Peterson.

Gleaves, A. (1928). The DeHaven Arctic expedition: A forgotten page in American naval history. *Proceedings, 305*(54), 579–591.

Hendrick, H. (1878). *Memoirs of Hans Hendrick, the Arctic Traveller, serving under Kane, Hayes, Hall and Nares, 1853–1876*. Trübner and Co.

Kane, E. K. (1854). *The U.S. Grinnell expedition in search of Sir John Franklin*. Harper & Brothers.

Kane, E. K. (1857). *Arctic explorations: The second Grinnell expedition in search of Sir John Franklin, 1853, '54, '55*. T. Nelson and Sons. There is no direct evidence of Kane meeting with Samuel Morton. However, given their proximity in Philadelphia, their common medical circles, Morton's membership alongside Kane's father in the APS, and Morton's influence on contemporary science, medicine, and ethnographic studies, it would have been unusual if such a meeting did not take place or that Kane would have come under Morton's influence (see, for example, O'Leary, Derek Kane. 2018. "The Arctic Other in American Cultural and Intellectual History." The Arctic Institute Center for Circumpolar Security Studies. Online at: https://www.thearcticinstitute.org/arctic-other-american-cultural-intellectual-history/ (Accessed October 2025)

Chapter 4
Isaac Israel Hayes, William H. Seward, and the First American Attempt to Purchase Greenland

Abstract By the middle of the nineteenth century, American writers such as Pliny Miles and explorers like Isaac Israel Hayes had provided the American reading public with volumes of information on Greenland and Greenlanders. It remained for the U.S. government, in the person of William Seward, to attempt to use this information to expand American territory north as it expanded west as well as into the Caribbean.

Keywords Greenland · Greenlandic Inuit · American route · Iceland · North Pole · Upernavik · Ellesmere Island · Cape York · Greenlanders · Kennedy Channel · William Seward · Alaska Purchase · Iceland · Cryolite

The resting place of Benjamin Franklin (the "Printer" who famously wrote his own epitaph to announce that he would return "In a new & more perfect Edition, Corrected and Amended By the Author") is located in the famous Christ Church Cemetery in the center of Philadelphia, itself the epicenter of American political and scientific life until the U.S. Civil War. It is only a few steps from the grave of Edwin De Haven, the U.S. Navy officer who can rightly hold the claim as the first American explorer of the Arctic. Franklin and De Haven are buried just 6 miles from the pharaonic tomb of Elisha Kent Kane along the Schuylkill River.

The other local polar explorer, Isaac Israel Hayes, is interred in a much more modest and, for its time, remote area. Hayes was only 22 years old when Kane was forced to amputate several of his toes after the disastrous winter at Van Rensselaer Bay in 1854. When he died in 1881, Hayes was buried in a small Quaker cemetery near West Chester, Pennsylvania, that even today is difficult to find. His fame lasted only briefly. There is one report of his grave being visited a year after his death on what was then known as Decoration Day, (now the American Memorial Day that honors those who served in America's military), but thereafter he lapsed into obscurity (Fig. 4.1).

Another American explorer of the far north passed in similarly modest ways. Pliny Miles was a tall, skinny, ebullient, brilliant wanderer and writer from Watertown, New York, who spent the mid-nineteenth century exploring the globe

P. J. Capelotti, *Lost Greenland*, Springer Polar Sciences,
https://doi.org/10.1007/978-3-032-27841-8_4

Fig. 4.1 The grave of Isaac Israel Hayes in April 2013 (photograph by the author)

and writing minutely detailed tracts on all manner of subjects, from steam travel to postal reform. As Elisha Kent Kane spun fantastic theories of mystical open seas around the North Pole, Miles in 1852 embarked on a serious, albeit completely impoverished tour of northern Europe, visiting, "the slightly out-of-the-way region of Iceland" (*New York Times*, 4 May 1865: 8). Unlike Kane, Miles' writing, as his brief obituary following his death in Malta in 1865 noted, depended "upon a laborious array of facts, rather than rhetorical effort" (*ibid.*).

Miles own description was to polish his original notes as little as possible. "The spirit of travel is the freshest at the time the travel is enjoyed; and all impressions are then the most vivid" (Miles, 1854: xiv–xv). And he took special aim at anyone who would impugn the character of the northerners who welcomed him. The popular Austrian travel writer, Ida Pfeiffer, had visited Iceland in 1845 and dismissed the people as greedy and unwelcoming. Miles, referring to Pfeiffer as 'Madame Trollope,' replied: "This is the woman that runs all over the world, and writes books about what she sees, and much that she does not see… then she pens all manner of false and libelous stories of the most kind, hospitable, unoffending race of people that the sun shines upon. The best comment that can be made on her book is, that she describes her journey to Mt. Hekla, and *ascent to the summit*, when the people here on the ground told me she never put her foot on the mountain at all!" (Miles *op cit.*: 123).

During his explorations of Iceland's culture, geography, and history, Miles studied that nation's relationship to its Greenland neighbor. His published account, entitled *Norðurfari; or, Rambles in Iceland* was well-received, and should have once and for all laid to rest multiple persistent notions about Greenland, Iceland, and the unique climate that bound them together. While the mystery of the missing Norse colony remained, Miles demolished any notion that explorers like Columbus would not have known of land to the west of Greenland.

> Iceland is not a myth, it is actual and real, a solid portion of the earth's surface… first discovered (by) a Norwegian pirate, in the year 860, almost one thousand years ago…. Greenland was discovered in 980, one hundred and twenty years after the discovery of Iceland…. The name of *Greenland* thus given to it, was as great a misnomer as *Iceland* applied to the neighboring isle. In reality, the two countries should change names; for Iceland is a country of green fields and fair flowers, while Greenland is covered with almost perpetual ice and snow… Eric the Red had a companion in his Greenland settlement, whose name was Heriulf. Biarni, the son of Heriulf, sailed from Iceland to join his father in Greenland, was driven south, and landed on the American coast—probably Labrador. Thus, the first discovery of America by Europeans was in the year 986, by Biarni Heriulfson, a native of Norway, though he sailed from Iceland. He returned north, landed in Greenland, and gave an account of his discovery. These facts I gathered from Icelandic genealogical tables; and all who have investigated the history of the northern nations, know with what accuracy these tables are compiled… Leif Ericsson having purchased the ship of Biarni Heriulfson, sailed from Greenland in the year 1000. The first land he made he called *Helluland*, or "land of broad flat stones." This was doubtless the coast of Newfoundland. The next coast he saw was covered with forest, and consequently he named it *Markland*, or "Woodland." This was probably Nova Scotia. The next land he discovered, still farther south, produced vines and grapes, and this he named "Vinland," a name the Icelanders ever afterwards used in speaking of the American Continent. We have the best of proof in their account of the climate and productions, in the length of the days, as well as in their maps and drawings, that their settlement was on some part of our New England coast, probably Massachusetts or Rhode Island. In subsequent voyages, these adventurous navigators sailed farther south; and it is supposed from the account they gave, that they proceeded as far as Virginia and the Carolinas. Timber, furs and grapes, were the most valuable articles the country produced; and for these, several voyages were made to Vinland, from Greenland, houses were built, and settlers resided in the country for at least three years; from 1011 to 1014. In their intercourse with the Indians, the Iceland and Greenland adventurers carried on their business about after the same political code that Raleigh, John Smith, and others, did afterwards. They first traded with the Indians, then fought them…. Voyages continued to be made to America, from both Greenland and Iceland, to as late a period as the middle of the fourteenth century. The last trip of which we have any record, is that of a vessel sent from Greenland to Markland (Nova Scotia) for timber and other articles. While returning, it encountered heavy storms, and was driven into port in the west of Iceland. The old Greenland settlements continued for a long period, the latest account we possess, coming down to the year 1484. When they perished, or from what cause, is unknown (Miles *op cit.*: 36–39).

Unfortunately for Miles, by the time his reasoned observations appeared, Kane's own first volume of Arctic travels had raced ahead and lodged itself into American popular culture. America was soon far down Kane's romantic American route to the North Pole, and that road ran through Greenland. It was as if northern spaces like Iceland and Greenland lost their Lost Colony and Open Polar Sea mysteries then so, in a sense, would the North Pole itself.

Regaining his health while witnessing the national mourning over the death of Kane—and no doubt poring minutely over the accounts of both Kane and Miles, Isaac Israel Hayes by the end of 1857 had repeatedly failed to convince any sponsors and scientific societies to back a new polar expedition via Greenland. But, as he wrote, "at twenty-five one is not easily discouraged" (Hayes, 1867: 4). He possessed both personal experience of the American route, so far as it existed in 1860, along with Kane's devotion to the open waters that would lead to the pole.

Lacking the wealthy connections of Kane, it took Hayes five years to raise the money required to fund his own Arctic expedition. With the mystery of the fate of the Franklin expedition largely settled by the Scotsman Rae, the main goal of his voyage would be to reach that other governing mystery of nineteenth century polar exploration: the open polar sea. Hayes wrote that he readily (and conveniently) accepted "the deductions of many learned physicists that the sea about the North Pole cannot be frozen [and] that an open area of varying extent must be found within the Ice-Belt which is known to invest it…" (Hayes *op cit.*: 1) (Fig. 4.2).

The overall goal to reach the open sea was couched, as all such expeditions typically had to be, in terms of the attainment of specific geographic knowledge. Once on the edge of that presumed open sea, Hayes sought nothing less than to complete a survey of the north coasts of both Greenland and what De Haven had first called 'Grinnell Land' (now seen as the central sector of Ellesmere Island), "and to make such explorations as I might find practicable in the direction of the North Pole" (*ibid.*). Even more than Kane, Hayes made it clear that an American expedition would try to reach the North Pole via Greenland.

Hayes also planned the now-requisite panoply of scientific studies including observations of air and water currents and temperatures, variations of gravity and magnetism, glaciology, natural history collections, and the mystery of the source of the Aurora Borealis. To accomplish this immense body of scientific work, Hayes would rely on himself and another of the men who tried in August of 1854 to leave the Kane expedition, Dr. August Sonntag, as well as an assistant astronomer, Henry Radcliffe.

It was not until the summer of 1860 that Hayes had could mount his expedition. He sailed from Boston in a small, refitted schooner, the *Spring Hill*—rechristened, on the eve of civil war, as the *United States*. The now twenty-eight-year-old was bound for the western shore of Kane Basin, where he was convinced from a sledge journey during Kane's expedition that he could reach the edge of the polar sea.

By the end of July, the *United States* was off Sukkertoppen and, with the wind holding steady, bypassed the southwestern Greenland settlements and finally dropped anchor in the harbor at Prøven in the Upernavik archipelago on 6 August, escorted by a fleet of indigenous hunters in kayaks. An ancient cannon under a Danish flag were the only symbols marking the colonial enterprise at the small sealing village whose name Hayes loosely translated as 'Experiment.' With rotting seal carcasses all around, Hayes considered the spot possessed little more than "a barbarous contempt for the human nose…" (*ibid.*: 32). Beyond the village, on the other hand, Hayes discovered a perfect Arctic valley, filled with flowers, mosses, and grasses.

It was in 'Experiment' that Hayes suffered his first of many setbacks. An official U.S. State Department letter sent to the Danish government requesting all assistance be offered to Hayes did not arrive for another year. Worse, disease had ravaged the

Fig. 4.2 Isaac Israel Hayes, in an image taken around the time of the American Civil War by Matthew Brady (Brady-Handy photograph collection, Library of Congress, Prints and Photographs Division)

local population of dogs from which Hayes had planned to purchase the necessary teams for his planned sledge work northward along the shores of Kane Basin. Local hunters' teams had themselves been decimated and no amount of trade goods could pry the remaining dogs loose from their owners.

Hayes left Prøven after 6 days, stopping at Upernavik where the *United States* moored next to a Danish trading brig, the first large vessel the expedition had seen since leaving Boston. During the first night in Upernavik, the expedition's boatswain and carpenter, a Gibson Caruthers of South Carolina, died in his sleep. His body was placed in the burial ground adjoining the church at Upernavik. Hayes directed a railing to be constructed around the grave, and an inscription to be placed over it. Caruthers became the third member of an American expedition to Greenland to lose his life, and the first to receive a formal burial there.

The loss of Caruthers was a catastrophe. He had been a member of De Haven's expedition and as such the only other expedition member besides Sonntag and Hayes himself who possessed Arctic experience. Hayes described his Greenland burial spot as "a sad place for human sepulture. It lies on a hillside above the town and is dreary and desolate past description" (*ibid.*: 38). Hayes received more bad news from the local administrator who informed him that ice had prevented the whaling fleet from passing beyond Melville Bay that summer. It was increasingly unlikely that the *United States* would be able to return even as far as Kane's Rensselaer Harbor and the observatory at Fern Rock, much less the coast of Ellesmere Island.

Hayes remained at Upernavik for 4 days, to bury Caruthers and to collect as much clothing as possible in the form of reindeer fur and seal and dog skins which, "placed in the hands of the native women [were] converted into suitable garments" (*ibid.*). He also recruited six Greenlanders to his cause, including one, Peter Jensen, who became his interpreter along with taking the position of dog-manager. This secured for Hayes Jensen's dog team, which necessitated another stop sixty miles north at a small hunting station.

The *United States* encountered heavy ice almost as soon as it exited the harbor at Upernavik, but within a few days entered clear water at the southern reaches of Melville Bay. At Cape York, Hayes went ashore to meet Hans Hendrik who had done so much to enable Kane's expedition and who had married and settled in the area. Hayes recorded that "six years' experience among the wild men of this barren coast had brought him to their level of filthy ugliness" (*ibid.*: 66). Hayes took Hendrik on board along with his wife and infant child. The expedition now numbered 23 souls.

Hayes wavered in the last weeks of August, alternately seeking an ice-free path up the west coast of Greenland while simultaneously looking longingly across Smith Sound at his preferred path north along the eastern shore of Ellesmere Island. In the end, a powerful storm finished his plans for either route when it pinned the *United States* at the entrance to Kane Basin for several days, forcing Hayes to seek shelter for the ship before it was destroyed. The storm blew for virtually the entire last week of August, and when Hayes departed the ship to climb a nearby cliff to gain a view north, the force of the wind was so great he had to steady himself against a rock for fear of being blown back over the cliff.

In the end, Hayes' only option was to run for protection in Hartstene Bay and begin the process of securing the ship for the winter. His predicament was dire. As he wrote, "the failure to cross [Smith] Sound was a serious disappointment... Besides ... my vessel was, apparently, so badly injured as to be unfit for any renewal of the attempt the next year" (*ibid.*: 91). Hayes named the shelter Port Foulke after a friend and sponsor from Philadelphia, William Foulke. In addition to being a lawyer, Foulke was an avid historian and geologist, who two years earlier had been credited with the discovery of the first complete dinosaur fossil in America (subsequently named *Hadrosaurus foulkii*). Hayes would dedicate his account of the expedition to Foulke.

A small fjord to the north Hayes also named for Foulke, and just to the north of it lay the Inughuit village of Etah. The spot was not advantageous for Hayes' plans, being some twenty miles south of where Kane's *Advance* was frozen in (and some eighty miles if one were to walk the shoreline to reach it).

The *United States* was made ready for the winter. The yards were taken down and the upper deck roofed over to make a house 8 feet high at its peak and 6 feet tall on the sides. Hunting parties filled the ship and a nearby storehouse with kills, and carcasses of more than a dozen reindeer were hung from the shrouds and rabbits and foxes suspended in clusters from the rigging.

The expedition set up an observatory on a terrace some 30 feet above the high tide mark. There, Sonntag "took general charge of such scientific work as we found ourselves able to manage" (*ibid.*: 93). These included a survey and sounding of the small harbor, magnetometry studies from an instrument set atop two kegs filled with beans and frozen in place with water. Temperature recordings were made three times a day and on every seventh day they were recorded hourly.

'Scientific' work also included the crew pilfering native graves and grave goods, finds rewarded by Hayes with "a glass of grog and the promise of other holidays..." (*ibid.*: 108). One such foray produced two Inughuit mummies that Hayes planned to contribute to a museum until Hans Hendrik's wife saw the bodies and recognized one as a close relative. Her fury forced Hayes to order the reinterment of the bodies, even after she was promised that Hayes could somehow reanimate the bodies, Frankenstein-like, once he had brought them to America. The episode was one of several that led Hayes to his appalling contention that the indigenous Greenlanders were "a very strange kind of people, and are an interesting study, even more so than my dogs, although they are not so useful; and then the dog can be controlled with a long whip and resolution, while the human animal cannot be controlled with anything. They might very properly be called a negative people, in every thing except their unreliability" (*ibid.*: 114).

Sledge work came to a halt when a disease spread throughout the dogs that by early January 1861 had killed all of them. The native Greenlander named Peter disappeared from the ship and his body was only located three months later. Hayes put the disappearance down to harassment of the boy by Hans Hendrik. Then the scientific programme came to a halt when Sonntag left the ship with Hendrik and both men vanished for over a month. Besides being the only trained scientist, Sonntag was Hayes' closest friend on the voyage and at 28 was his exact age as well.

When Hendrik finally returned to the ship in late January, he was alone. Sonntag was dead after a fall through the ice. Hayes was devastated and never forgave Hendrik for what Hayes saw as either incompetence and/or actual malice in both the second and third expedition deaths. Hayes had Sonntag's body recovered and reburied in the observatory. No similar efforts were made to recover the Greenlander Peter's body.

Without Sonntag and reduced to whatever dog teams he could scrape together from trading with local Greenlanders, Hayes now tried to salvage anything he could from the expedition. First and foremost, he sought to solve the mystery of the open polar sea once and for all. His plan remained the same: to cross Kane Basin to Ellesmere Island and try to reach the shores of the polar sea.

After a brief return to Van Rensselaer Harbor where he found no trace of the abandoned *Advance* and revisited the Fern Rock observatory where the dead of Kane's expedition has been left, Hayes set out in early April with every member of his surviving crew to cross Kane Basin by sledge. It was during this crossing of the body of water north of Smith Sound that Hayes proposed that it be named "for the leader of the expedition which first defined its boundaries—I mean, of course, Dr. Kane" (*ibid.*: 310).

Halfway across the basin, Hayes realized that his large party was getting nowhere. They had advanced 30 miles from the coast of Greenland but by Hayes' estimate had traveled five times that distance in their zig-zag courses. At the end of April, he sent most men back to the ship and continued on with just three others. It was not until May 11 that his small party reached the opposite shore, after thirty-one back-breaking days that covered little more than 2 miles per day. Hayes made camp at the same farthest latitude north he had reached during the Kane expedition. From there, every step would be in new Arctic lands.

When his strongest remaining crewmember broke down, Hayes left him with a second man and then continued north with his personal secretary, George F. Knorr. "My purpose now was to make the best push I could, and … reach the highest attainable latitude and secure such a point of observation as would enable me to form a definite opinion respecting the sea before me…" Hayes saw in the distance "increasing darkness and … water-sky to the northeast" (*ibid.*: 344).

On May 18, 1861, Hayes reached the northern terminus of his efforts. Ascending to a cliff he reckoned was about 800 feet above the sea, Hayes looked out over a polar ocean filled with loose ice, arrayed so as to appear as a magical road leading north. It was "as the delta of some mighty river discharging into the ocean, and under a water-sky, which hung upon the northern and eastern horizon, it was lost in the open sea… Suffice it to say that all the evidences showed that I stood upon the shores of the Polar Sea…" (*ibid.*: 349).

Hayes' observations appeared to support those of Kane, but they were just as confounding. His estimate of his position was 81°35' North and 70°30' West which, if accurate, would have placed him not on a cape on the edge of a polar sea but a hundred miles into the interior of Ellesmere Island. Similarly, Hayes was wrong in his dismissal of Kane's maps. He incorrectly placed an island in the middle of Kennedy Channel and misplaced the position of the Humboldt Glacier as a result. He later placed the burden of his navigation errors on the Smithsonian Institution, which had taken his scientific notes to publish, and five years later still had not done so and, Hayes claimed, had not given him access to his original notes to write his own book (Fig. 4.3).

Despite these mistakes, Hayes was under no illusion as to his true position, writing that "all the evidences showed that I stood upon the shores of the Polar Basin, and that the broad ocean lay at my feet: that the land upon which I stood, culminating in the distant cape before me, was but a point of land projecting far into it…"

Fig. 4.3 Open water flows around scattered ice floes in a drawing based on Hayes' sketch of '*The Shores of the Polar Sea.*' From Hayes' *The Open Polar Sea* (1867)

(*ibid.*: 349–350). In truth, Hayes was standing not on the shores of an open polar sea but on the west side of Kennedy Channel, and the open water he spied merely the annual current-assisted break-up of the ice in the area.

This perpetuation of Kane's view of a broad open sea around the North Pole would have serious consequences for polar expeditions for the next 30 years. As for Hayes, he celebrated his apparent discovery with the thought that he would soon return, to sail upon "these ice-girdled waters [that] might lash the shores of distant lands where dwell human beings of an unknown race…" (*ibid.*: 352).

Hayes retreated to the *United States* and in mid-July of 1861 the ship was freed from the ice and began its voyage home. A month later the ship dropped anchor at Upernavik. There Hayes learned of the inauguration of a new president, Abraham Lincoln, and the subsequent beginning of the American Civil War.

By the time Hayes had returned to Boston in October, his dramatic tale of men dying in the Arctic in pursuit of an obscure geographic mystery was suddenly and completely irrelevant. Scanning a Boston newspaper with its appalling descriptions of wartime carnage, Hayes returned to the ship, as he wrote, "sad and dejected, and found my way on board again through the dull, dull fog" (*ibid.*: 450).

Hayes immediately volunteered his services as a doctor and was placed in charge of the 4500-bed Satterlee General Hospital in Philadelphia, where the number of patients swelled to over 6000 after the Battle of Gettysburg in July of 1863. Hayes' skill and leadership was credited with the loss of only 260 patients at a time when access to medicines and sanitation were problematic at best.

Hayes did not publish his account of the *United States* expedition until 1867, two years after the war's end and six years after the return home of his small schooner. His navigation led the German cartographer August Petermann to name the coast of Greenland from Melville Bay to Kane Basin 'Hayes Peninsula.' The northwest shore of Greenland was now marked as an American preserve. The Kennedy Channel was now flanked by Grinnell Land on the North American side and Hayes Peninsula on the Greenlandic. It flowed past Kane's Rensselaer Bay and Fern Rock, past capes named for Andrew Jackson, Thomas Jefferson, and the U.S. Constitution, then past Washington Land, all the way to the Lincoln Sea, which Hayes had done nothing to dispel as an open body of water leading to the North Pole.

One individual with a keen appreciation for the seeming geographic niceties Hayes espoused was also the one individual with the power to claim them once and for all for the United States. This would be Abraham Lincoln's Secretary of State, William Henry Seward, who had just pushed through the purchase of Alaska from Russia and begun the process of acquiring the Pacific kingdom of Hawai'i, the Panama Canal Zone and, from Denmark, the Virgin Islands of the Caribbean. As his biographer noted: "Although "empire" is today generally not considered a positive term, Seward used it often, and he would be proud today, as he should be, for his central role in founding the American empire" (Stahr, 2012: 547). Just as Hayes in 1867 was publishing his account of wandering the depressing streets of Boston upon the return of his expedition in 1861, the American flag was raised over Sitka, Alaska.

William Seward then turned his land lust towards two other Danish possessions, Iceland and Greenland. There was urgency in Seward's expansionism. What had been known as British America became the Dominion of Canada on July 1, 1867, with its external relations governed from London. Plans for a rail line from Nova Scotia to British Columbia to connect the Atlantic to the Pacific would effectively bypass the need for any British Empire goods to be shipped to or transshipped through the United States (the final spike that connected British Columbia to the Atlantic via the Canadian Pacific Railway was in fact driven into the ground at Craigellachie, British Columbia, less than two decades later, on November 7, 1885). With a fractured United States in the middle of a badly botched Reconstruction from the Civil War, and just at the outset of a protracted series of Plains Indian Wars, it was not difficult to imagine that the economic engine of North America would be Canada, not America.

Not everyone was on board with Seward. Congress was reluctant to appropriate the funds for Seward's vast land purchases, and newspapers around the country all but mocked his propensity for adding new territories to a nation struggling to regain its footing after a disastrous internal war. A Connecticut newspaper said that if the Secretary of State was not stopped, he would "buy up the whole hemisphere from the glaciers of Greenland to the volcanoes of Terra del Fuego" while another predicted that new members of Congress would soon be seated from "Hawaii, Fiji, Madagascar … and perhaps Iceland and Greenland"(Stahr *op cit.*: 496).

Others were just as adamant in their support of Seward's expansionist agenda, especially as they stood to profit from it. One such enthusiastic supporter was former Treasury Secretary Robert J. Walker. Like Elisha Kent Kane, Walker had graduated at the top of his class at the University of Pennsylvania. He married a great-granddaughter of Benjamin Franklin whose brother would become a future director the U.S. Coast Survey. A long-time Democratic lawyer and Washington fixer, Walker had pocketed a cool $20,000 in fees for his legal counsel during the Alaska purchase. As they discussed his treaty with Denmark for the acquisition of the Caribbean islands of St. Thomas and St. John in the summer of 1867, Walker suggested to Seward "the propriety of obtaining from the same power Greenland, and probably Iceland also" (quoted in Pierce, 1868: 1).

Seward never turned down the chance to acquire new lands for the United States and asked Walker to send him a detailed *aide-memoire* to supply him with compelling reasons why he should move ahead. Walker knew just the people to produce such a report. He contacted an acquaintance in the Coast Survey who in turn hired a young geologist to write a comprehensive survey of the resources of both Iceland and Greenland.

The young geologist, Benjamin Mills Pierce, spent several months scouring available sources (although not following in the wake of Pliny Miles to actually visit either place himself). Before the end of 1867, Pierce turned in a fine 72-page report directly to the Superintendent of the United States Coast Survey, who also happened to be his father, and also named Benjamin Pierce. The Superintendent forwarded the report to Walker, who in turned passed it on to Seward with his compliments and a recommendation.

> I have heretofore expressed the opinion that we should purchase Iceland and Greenland, *but especially the latter* (italics mine). The reasons are political and commercial. The proof has heretofore been submitted by me, that the government, recently partially established in British America, called the Dominion of Canada, was gotten up in England in a spirit of bitter hostility to the United States. It then was, and still is, intended to embrace all British America, extending from the Atlantic to the Pacific, with a railroad from Halifax to Puget sound, and an area exceeding that of the United States prior to the purchase of Alaska. By this great purchase, we have flanked British America on the Arctic and Pacific, cutting her off entirely from the latter ocean from north latitude 54° 40' to 72°, leaving the new dominion but 5° 40' on the Pacific, pressed between Alaska on the north and California, Oregon, and Washington Territory on the south, with even British Columbia now being rapidly Americanized. Now, the acquisition of Greenland will flank British America for thousands of miles on the north and west, and greatly increase her inducements, peacefully and cheerfully, to become a part of the American Union (Walker quoted in Pierce, 1868: 3–4).

Walker did not expect the U.S. Secretary of State to read each and every footnote of Pierce's comprehensive report, so he took the liberty of abstracting it, calling attention to the necessity of securing Greenland's cryolite mines, its fisheries, and its foothold on what American explorers reliably informed him was an open polar sea, which could be profitably explored using the balloon reconnaissance technologies so recently employed during the Civil War:

> "This is the largest island in the world. It extends, according to Petermann, (a very high authority,) from longitude 20 west of London to 175 east, thus passing nearly half round the globe. Its area, thus elongated, would be about 1,800,000 square miles, or largely more than half the size of all Europe, but with a far greater shore line. Not a hundredth part of this vast

region has been explored, but the geologic structure indicates great mineral wealth. Greenland extends, according to Petermann, from north latitude 59o 57' 30" to within 50 miles of the North pole, with a length of several thousand miles. The open Polar sea of our American explorers is regarded by Petermann as only a large bay, north of which the land closes again. This may be, although it would still leave a smaller Polar sea; but in view of the discoveries of Wrangel, and still later of an American captain in the Arctic ocean, north of Behring's straits, is it not probable that a portion of the land elongated by Petermann, west of our supposed Polar sea, may not be continuous, but, as suggested by General T.L. Kane, a congeries of islands, (somewhat resembling the Aleutian group,) still constituting a part of Greenland, but leaving probable openings between these islands to the Polar sea and the North pole? In view, also, of the mild temperature at Behring's straits, and the fact that Parry ascended to latitude 82o 47' unobstructed by land and with none in view, is not the proper route by these straits for our next expedition to the North pole, and should not balloons similar to those used in military reconnaissances be elevated when necessary to descry distant objects?

"The shores of Greenland much more than those of any other country are indented with deep bays, inlets, estuaries, and fiords, some of them possibly extending from the western to the eastern coast, presenting an immense shore line, and furnishing most extensive and protected fishing grounds. "These inlets are bordered by meadow lands, beech and willow, whence the name of Greenland." The population, neglected by Denmark, lives mainly by hunting, including furs and the fisheries.

"Greenland has in vast quantities whale, walrus, seal, and shark, cod, ivory-cod, salmon, salmon-trout, and herring; foxes, wolf, reindeer, bear, hare, myriads of birds, including the king duck, eider duck, dorskin, petrel, gull, brent, burgomaster, goose, killiwoke, lolard, and sea swallow, &c, &c. Good coal is found on the western coast at various points, extending far north, most cheaply mined, and close to good harbors. From the best of these northern harbors of Greenland there is believed to be practicable summer ocean steam navigation 1,500 miles to Alaska, extending, also, through Behring's straits to China or Japan, or southward to Sitka, Puget sound, the Oregon river, San Francisco, &c., &c. This Greenland and Alaska coal may possibly render this transit practicable, and would be of immense value in connection with the fisheries.

"The whale fisheries of Greenland, in 1864, were of the value of $400,000. The climate of south Greenland is one of "unusual healthfulness, and clear atmosphere." The limits of summer are from May to September. All the shores and inlets of Greenland abound with animal life, such as fish, birds, &c. Kane states the all-important fact of *a vast increase of animal life as you approach the most northern arctic waters*. The rocks and geology of Greenland, as before stated, besides the valuable coal discovered, indicate vast mineral wealth. Kryolite, a most important mineral, and of very rapidly increasing use and value, is found *only* in Greenland. One of the mines is 80 feet thick. This mineral is mined in large quantities, its rapid development being due in great part to American enterprise. It is used in the manufacture of soap, soda, and soda salts, and yields a residue which has a great value as a flux in the treatment of difficult metallic ores. It also makes a fine glass, and has been employed in the manufacture of *aluminium*, and its alloys, a most valuable metal, being very brilliant, one-third the weight of zinc and one-fourth that of silver, good for coins, much used in France, one-third the price of silver, valuable for jewelry, tenacity equal to steel, valuable for watch cases, mirrors, spectacle cases, opera and field-glass cases, pendulum rods and small weights and balances, instruments of precision and where great lightness is required, spoons, forks, dinner service, cooking apparatus, being unaltered by water, vinegar, salt, and other organic matter. The mines appear to be inexhaustible, and are of great and rapidly increasing value. Geologists all look for *new and immense* mineral developments in Greenland. Kane's book is most valuable. He found "fabulous numbers of whales in Whale sound; swarming also with sea animals and myriads of birds." Hayes describes "the green meadows there as a paradise, with swarms of whales, walruses, &c.,

and in adjacent seas." On the extreme north coast the north winter winds bring *mild weather*, because they come, as believed, from what the Russians call Polynya, referring, in some way, to a Polar sea. Wrangel observed that the northwest winds, as well as some of the northeast, brought with them a thick, moist fog, so that the clothes and tents were wet through.

"These are the main facts as to Greenland and Iceland, taken chiefly from Mr. Peirce's great report, which I consider as a most valuable contribution to science...

"The same glacial action which has cut up Iceland into so many inlets and fiords has, so far as explanations have been made, produced similar results, on a much larger scale, in Greenland, thus probably rendering its shore-line nearly equal to that of all the United States before the purchase of Alaska. Hence, Greenland has immense fisheries, most of which are undeveloped. But all our explorers attest the important fact that, just as you proceed north in the Arctic towards the pole, the profusion of animal life, including birds and fishes, *is wonderfully increased*. Now, the nation with such vast fisheries must not only have the largest commercial marine, but the best and greatest number of seamen, and, as a consequence, enabled promptly, when required by any emergency, to put in operation the largest and most effective navy. Such vast fisheries and extensive coasts and numerous harbors, especially with abundant good coal there, must greatly antedate the period when the United States will command the commerce of the world...

"The same philosophic reasons, growing out of the far warmer temperature of the Arctic ocean, as compared with the land, which indicate a Polar sea at the north, would lead to the conclusion that a similar Polar sea exists at the South pole, and that in the Northern and Polar seas will be found far the greatest fisheries of the globe, and exhaustless reservoirs of food for man" (Walker, quoted in Pierce, 1868: 2–5).

Seward needed little more encouragement, but events stymied his Greenland ambitions. The acquisition of Greenland all but required a simultaneous absorption of British Columbia into the American Union in order to connect the American Pacific Northwest to the newly acquired Alaska. During the earlier debate in the U.S. House of Representatives over the appropriations for the Alaska purchase, one Congressman in opposition to the funds triggered bemused laughter when he quipped that if the House approved this legislation, Seward would be back for more money to buy Iceland and Greenland (Stahr *op cit.*: 516).

But Rep. Cadwallader Washburn from Wisconsin's 6th District was not kidding. He pulled out several pages of Pierce's report, as yet unpublished. A clerk at the State Department had apparently leaked the Greenland/Iceland report to a journalist, who had in turn leaked the report to Washburn. It required furious lobbying by Walker—along with rumored bribes to several influential politicians (including the incorruptible Thaddeus Stevens, who died before he could receive his), to say nothing of a suggestion that Seward had perjured himself before Congress in denying any knowledge of bribery, before the Alaska appropriation passed both houses of Congress and was signed into law by Andrew Johnson on 27 July 1868 (Stahr *op cit.*: 517–518). The United States had acquired Alaska, but the embarrassing drama had lost America any chance to add Greenland, or Iceland, to the Union any time soon.

At this point, Greenland should have receded back into northern mists, but it was kept in the American consciousness by a man who himself seemed to have appeared out of a fog. Charles Francis Hall was born in New Hampshire. Or it might have been Vermont. He was raised in Rochester, New York, probably. What is certain is that he arrived in Cincinnati, Ohio in 1849, with a young wife and no money and attempted to make a living as a journalist.

Hall became a newspaper publisher at a time when brief items of news or rumor or gossip about the fate of Sir John Franklin appeared with regularity. A decade later, after the Greenland expeditions of the internationally famous and well-connected and supremely educated Elisha Kent Kane had failed to solve this greatest of Arctic mysteries, and left behind only tantalizing clues about an open Arctic sea, the impoverished Hall, without formal education or means or connections of any kind, left his wife and family in Cincinnati and set out on a mission to find Sir John and, with a bit of luck, the North Pole. Nearly successful in both endeavors, Charles Francis Hall would in the process become the greatest Arctic anthropologist in American history, his bones interred on an impossibly remote northern shore of Greenland, in the northernmost grave of any American.

References

Hayes, I. I. (1867). *The open Polar Sea: A narrative of a voyage of discovery towards the North Pole in the schooner United States*. Hurd and Houghton.

Miles, P. (1854). *Norðurfari; or, rambles in Iceland.* Charles B. Norton.

New York Times. (1865). *Death of Pliny Miles, The American Advocate of Cheap Postage*, May 4, 1865 (8). See also: Borgford, Brian. 2024. Nineteenth Century Education in/of Iceland: A journey of learning for Pliny Miles—1852. *Icelandic Roots*, Retrieved October 28, 2025: https://www.icelandicroots.com/post/nineteenth-century-exploration-in-of-iceland-a-journey-of-learning-for-pliny-miles-1852

Pierce, B. M. (1868). *A report on the resources of Iceland and Greenland.* U.S. State Department.

Stahr, W. (2012). *Seward: Lincoln's indispensable man.* Simon & Schuster.

Chapter 5
Charles Francis Hall, Adolphus Greely, and the Northernmost Burial of an American Explorer

Abstract By the 1870s, Northwest Greenland was firmly established as the place where American explorers needed to reach if they were to have any chance to become the first human to stand at the North Pole. The earlier expeditions of Kane and Hayes were furthermore used to justify the American acquisition of Greenland as an Arctic colony and resource base.

Keywords Greenland · Greenlandic Inuit · North Pole · American route · Cryolite · Frobisher · Upernavik · Kennedy Channel · Etah · Fort Conger · 1868 State Department Report on Greenland

> *We hope, in conclusion, that it is not out of place to quote the following remark from Porter's Progress of the Nation. It may have a bearing upon the question of Iceland and Greenland: The negative advantages offered to a state by the possession of its colonies consist in this, that their power and resources cannot be rendered available against it.*
>
> —Benjamin Mills Pierce, 'Advantages of a colony,' from: *A Report on the Resources of Iceland and Greenland* (p. 51).

In the conclusions to his report on the resources of Iceland and Greenland, Benjamin Mills Pierce stressed that nothing less than national honor demanded the continuation of the explorations of Greenland and the Arctic Ocean begun by Kane and Hayes. "The United States has inaugurated a brilliant line of investigation by Kane's and Hayes's expeditions. Let us hope, for her own honor and for the fame of American navigators, that what she has so gloriously begun will be followed up. If national glory has any meaning in the present state of civilization, it can be gained in no nobler way than by such achievements" (Pierce, 1868: 51).

Pierce also made a case for the direct American acquisition of both areas. The primary rationale was the immense resource base Greenland especially possessed. Greenlandic cryolite was already mined and shipped to the foundries of Pittsburgh, while the massive fisheries were well known. Yet even Pierce could have suspected in the mid-nineteenth century how important the scramble for Greenland would become to the twenty-first century. Pierce's arguments are as familiar in 2025 as

P. J. Capelotti, *Lost Greenland*, Springer Polar Sciences,
https://doi.org/10.1007/978-3-032-27841-8_5

they were as he wrote his report in 1867. The Unites States required Greenland so the Greenland could not be used against the United States. The only difference in the two eras was the opposing superpower in question. For Pierce it was Great Britain.

Missing from Pierce's resource calculations—as, one could argue—they are missing from similar discussions in 2025—was any consideration of the greatest resource in Greenland, the Greenlanders themselves. In the few places where Pierce even mentions them, he approvingly quotes Petermann approvingly quoting Hayes: "The Esquimaux are a shiftless people, living as they do from hand to mouth, without laying up provision for the future. Indeed, the Esquimaux would die out even in Germany, as they would find no nourishment in winter, (without laying it up during the summer,) as they certainly do find food in the seas of these regions" (Pierce *op cit.*: 59–60). Apparently, the survival for millennia on the world's most inhospitable shore counted for little if one did not possess the requisite skills to live through a January in Berlin.

One can hardly fault Pierce his oversight, if not his prejudice. At the same moment as William Seward eyed Greenland, others had found vast riches much closer to home, on the Western Plains. Unlike Greenland, where the occasional American explorer turned up starving, incompetent, and ultimately dependent on Greenlandic knowledge for their survival, the rapacious speculators who overwhelmed the Dakota Territory in the aftermath of the Civil War had the advantage of a suddenly out of work U.S. Army and soon, a dedicated railway, to assist them.

John Gregory Bourke, a young captain and Civil War veteran from Philadelphia, served in the Western army as aide-de-camp of perhaps the greatest American Indian fighter, General George Crook. Bourke witnessed firsthand the Plains Indians Wars, from the Apache Wars after 1871 to the Great Sioux War of 1876 to the final capture of Geronimo in 1886. In Bourke's view, these remorseless conflicts over land and the resources underneath them could have been avoided had only the U.S. honored its treaty commitments and mandated that the encroaching whites fairly share the gigantic wealth they were determined to hoard to themselves.

> Much of our trouble with these tribes could have been averted, had we shown what would appear to them as a spirit of justice and fair dealing in this negotiation. It is hard to make the average savage comprehend why it is that as soon as his reservation is found to amount to anything he must leave and give up to the white man. Why should not Indians be permitted to hold mining or any other kind of land? The whites could mine on shares or on a royalty, and the Indians would soon become workers in the bowels of the earth. The right to own and work mines was conceded to the Indians by the Crown of Spain, and the result was beneficial to both races... The policy of the American people has been to vagabondize the Indian, and throttle every ambition he may have for his own Elevation... (Bourke, 1891: 244).

Bourke's desultory conclusions based on his Western experience were not shared by those like Kane and Hayes who sought to use Greenlandic resources and local knowledge to advance their own ambitions. Hayes' demoralized return to Civil War America had done little to dissuade his desire return to the Arctic. But when he finally got the chance, it was as a tourist, not the leader of a polar expedition.

By the time Hayes published his book in 1867, he faced competition from his remarkable contemporary, the self-made explorer Charles Francis Hall. Gathering support from Henry Grinnell and from a whaling company in New London, Connecticut, Hall was determined to sail north to solve the mystery of Franklin and the Open Polar Sea by using the one source of data that Hayes had endlessly disparaged. It was the same source of legitimacy that Bourke had urged on all dealings between Americans and the array of diverse peoples an American empire would inevitably encounter: the primary acknowledgement of the value of Indigenous knowledge and the primacy of Indigenous land rights (Fig. 5.1).

In the spring of 1860, Hall arrived in New London, Connecticut with a few bits of expedition gear, multiple debts, and the zeal of a man on a mission. The staff of the 'New Franklin Research Expedition' consisted of Hall and Hall alone. He boarded one of Henry Grinnell's whalers as a passenger on a routine search for whales in Baffin Bay, exuding the self-confidence of an American Midwestern hayseed with a messianic complex. "Everything related to the Arctic Zone is deeply interesting to me. I love the snows, the ices… I love the circling sun, the long day, *the Arctic night when, when the Soul can commune with God in silent and reverential awe*! I am on a mission of love. I feel to be in performance of a duty I owe to mankind, myself, God!" (Loomis, 1971: 72).

It was on the shores of Baffin Bay that Hall first met the Inuit and was soon firm friends with an Inuit couple named Ebierbing (nicknamed 'Joe') and Tookoolito (nicknamed 'Hannah')—both had been to England and spoke English. It was this

Fig. 5.1 The only known photograph of Charles Francis Hall, from 1870 (public domain via Wikimedia)

couple that enabled Hall's first foray into the Arctic to be remarkably productive. A true anthropological explorer, Hall began to learn the Inuit language, corrected Frobisher Strait to the bay that it is and then, through local knowledge, uncovered both native history and the historical archaeology of the expeditions of the British explorer Martin Frobisher three hundred years earlier. Forced by ice conditions to remain in the north for a second winter, Hall took the opportunity to scout even further afield.

As with Hayes a year earlier, Hall returned to the United States in 1862 and found a country at war with itself and with no interest in an abstract subject like polar exploration. It was two years before Hall was able to return to the Arctic, hopscotching north on board several whaling ships of opportunity. It was fully five years more before Hall finally reached his intended destination of King William Island. There, using his educated layman's experience in recording both history and archaeology, he expected to learn more of Franklin's fate from native testimony and from any artifacts that might lie around the island.

At Shepherd Bay, opposite King William Island on Boothia Peninsula, Hall found what he was looking for: a cache of Franklin relics that he immediately traded for. In addition to the artifacts, Hall recorded native testimonies about a trail of British bodies that lay exposed across King William Island. Through these oral histories, Hall reckoned that he had accounted for nearly eighty of Franklin's lost officers and men.

Hall remained in the Arctic for years. He returned to the U.S. only in the late summer of 1869. Through Inuit oral testimonies, he had demonstrated more or less conclusively that Franklin and his crew had died fairly quickly—even as they clung to the material anchors of their Victorian society. He had also proved that it was possible to survive for many years in the High North by adopting and adapting to an Inuit way of life. With the Civil War over, Alaska now America, and the U.S. in search of even more frontier lands, Hall now intended to exploit his knowledge of northern life in a journey to the North Pole itself via the American route along the northwestern coast of Greenland.

Hall was aided by a renewed post-war fascination with Arctic exploration, an interest fueled, ironically, by a nationwide series of lectures by his rival along the American Route, Isaac Israel Hayes. Hayes was urging a return to the north as he sold copies of his new book. He also signed on to a tourist cruise of Greenland that allowed the artist William Bradford to experience the vistas of ice and mountains in both Greenland and Labrador. Bradford's subsequent dramatic landscapes contributed to a new American vision of a romantic Arctic (Fig. 5.2).

But it was Hall's decade of experience in the ice, his knowledge of the languages and cultures of the Far North, along with the renewed public interest in the Arctic, that earned him the patronage of a sympathetic Secretary of the Navy, George M. Robeson. This enabled him to secure a titanic government grant of $50,000, an unheard-of sum for a private citizen. Even more remarkably, he also secured command of a U.S.-sponsored vessel, a large tugboat called the *Periwinkle*. This vessel was reconditioned as an Arctic cruiser and renamed *Polaris* and given over to Hall

Fig. 5.2 *Voyage of the Polaris* (1875), by William Bradford (public domain via Wikimedia)

with the express mission of reaching the North Pole. While most of the officers were American, much of the crew was made up of German immigrants.

At the start of this expedition, in the summer of 1871, Hall's fabulous run of luck continued. In southwestern Greenland, he met with a hydrographic expedition that informed him of excellent ice conditions in the far north, so Hall ordered *Polaris* to steam directly for Smith Sound. At Upernavik, he was able to secure the services of Hans Hendrik, the same Greenland hunter who had figured so prominently in the voyages of both Kane and Hayes. And unlike the schooners of Kane and Hayes, Hall's steamer *Polaris* breezed through Smith Sound and into Kane Basin.

The *Polaris* passed the wintering spot of Hayes' *United States* late in the afternoon of 27 August and just three hours later passed Kane's Van Renssalaer Harbor. The steamer even found an ice-free Kennedy Channel and eventually entered a smaller basin to the north—unseen by either Kane or Hayes—which Hall proudly and justifiably named after himself (Fig. 5.3).

The Americanization of the American Route to the North Pole continued. An icy strait north of Hall Basin took on the name of his government patron Robeson, then serving as Ulysses S. Grant's Secretary of the Navy and the man who had successfully lobbied Congress to support Hall's expedition with its grant of $50,000. Beyond Robeson Strait lay Hall's Lincoln Sea, and the shores of the Arctic Ocean. It would be another forty years before Robert Peary seriously probed that ocean and reported nothing but ice between the shores of the Lincoln Sea and the final miles approaching the North Pole.

Polaris was swept up in the ice and pushed south for a few miles as Hall and the captain of the vessel, Sidney Budington, searched for a winter anchorage. They eventually retreated behind a large, grounded iceberg and blocked the ship in for the winter. The anchorage was named 'Thank God Harbor.' Before the slight retreat

Fig. 5.3 The *Polaris* and *Congress* at Godhavn, Disco Island off the coast of Greenland (from a magazine article from May 1873)

southwards, *Polaris* recorded a then-record for a ship at 82° 11' North. It was an astounding accomplishment for a novice mariner and flew in the face of every bit of experience of the first three American Arctic expeditions under De Haven, Kane, and Hayes.

Hall and the *Polaris* had reached the very northern shores of Greenland, little more than 450 miles from the North Pole. To the north Hall witnessed the sight that had eluded all previous American expeditions. Instead of some mythical open polar sea, he confronted the reality of unbroken ice pack between the shores of Greenland, Ellesmere Island and, if he could have scouted that far, the North Pole. There was no land bridge to the Pole. Neither was there any tropical sea with an unknown race cavorting on the rim of an active volcano. And, at long last, for the many searchers and speculators of the previous quarter century, there was no sight of *Erebus* and *Terror* coursing across an ice-free sea in search of a passage south.

As an experienced Arctic traveler, Hall was anything but deterred by this new reality. To the contrary, he was poised to take to the ice and strike for the North Pole in the spring of 1872. As he had written in late August, the "prospects of the expedition are fine; the weather beautiful, clear, and exceptionally warm… Our coal bunkers are not only full, but we have full ten tons yet on deck… Never was an Arctic expedition more completely fitted out than this" (Loomis *op cit*.: 271).

The attainment of the Pole itself might have been beyond Hall's grasp, but given his starting point there is little question he would have made a serious probe northward. On a mid-October survey, Hall ascended a mountain and gained a clear view of the surrounding lands no Westerner had seen before. Across Robeson Strait, Ellesmere Island trailed off to the west. The northern Greenland territory Hall spied did the same towards the east. The icy ocean and the absence of any visible land to the north meant that Hall could start his attempt without having to burden himself with geographic fantasies like polar land bridges and open seas. It seemed apparent that both Ellesmere Island and Greenland were just that, enormous islands that ended at just about the exact place where Hall stood. The Lincoln Sea was covered by a permanent, shifting pack of ice.

What happened next went unexplained for nearly a century. After a series of brief sledge excursions in the fall to test his gear and the crew, Hall set out on 10 October for an extended trip north. When he returned to the ship two weeks later, he drank some coffee—crew members later differed on who prepared the coffee and exactly which pot Hall got his from—and was taken suddenly and violently ill. In moments of clarity over the next week or so, he complained that he was being poisoned by, among others, Emil Bessels, the Smithsonian Institution scientist on board.

Charles Francis Hall died in agonizing pain, still complaining about plots against him, on 8 November 1871. He was only 50 years old. Or perhaps 49. Maybe 51. No one knows the exact moment of his birth beyond 'circa 1821.' He had written out prodigious notes on his expedition and these, with all his other papers and books, were misplaced and lost—whether by accident or design—when the expedition tried to retreat a year later. He departed polar exploration as mysteriously as he had joined it a decade earlier.

Hall was interred in a stony grave on the shores of Thank God Harbor, northwest Greenland. No American before or since has been buried in the earth so far north (Fig. 5.4).

With Hall's death, the ship's captain Sidney Budington ascended to leadership of the expedition, at which point the entire effort quickly fell apart. Discipline vanished as the ship's officers lost all authority, especially with the largely German crew. Budington himself retreated to his cabin to partake of the ship's alcohol stores. Half-hearted polar attempts were made by small boat the following June, but the navigator of the expedition recalled afterwards that with Hall's death any serious thought of the North Pole was abandoned.

When the *Polaris* broke free from the ice on 20 June 1872, Budington at once realized the damage the ship had suffered over the winter. Mechanical pumps were set to working to keep the ship afloat. As the ship explored a small lead in mid-August and began the long and dangerous retreat to the south, it was quickly and fatally caught in the ice. For two months the vessel drifted south, first in the middle of the Kennedy Channel and then Kane Basin, always with land out of reach many miles away to either side of the helpless ship.

On 12 October, drifting into Smith Sound, a gale blew up and shattered the ice around the ship and threatened to sink her. In the ensuing panic, the surviving expedition members were split into two groups. One, made up of ten crew and all of the Greenlanders, was marooned on an ice floe and began a long drift south. Only the hunting skills of the Greenlanders kept this party alive until it was rescued some six months later off the coast of Labrador. As extraordinary as this drift assuredly was, it was an almost identical repeat of a German catastrophe off the east coast of Greenland just four years earlier, when the polar expedition vessel *Hansa* was crushed and sunk and its crew marooned on the ice until rescue the following May.

Fig. 5.4 Tablet erected on the grave of Captain C.F. Hall of the USS *Polaris* by British polar expedition, 1875

The fourteen crew that remained on board *Polaris* drove the foundering ship ashore near Etah. It was somewhere between the frantic scramble onto the ice floe and the ship's grounding when Hall's archive of papers, books, and scientific instruments, was lost. The *Polaris* party built a shelter with wood from the ship, survived the winter with the help of the Etah Greenlanders, and the following spring constructed two boats from the bones of the ship and sailed south where they were picked up and taken to Scotland and eventually returned to the United States.

Navy secretary Robeson convened a board of inquiry in Washington as the two disparate groups of survivors returned. The resulting testimony was conflicted and contradictory on virtually every major question. Neither could the board examine Hall's papers or even his body, leading it to settle blame upon everyone and no one. As for the question of Hall's death, forensic data would not be forthcoming for nearly a century.

In 1968, a Dartmouth College professor of English and biographer of Hall's by the name of Chauncey Loomis received permission to exhume Hall's body from its grave at Thank God Harbor. A hasty autopsy of the frozen body returned some small clippings of Hall's hair and fingernails to be tested in the U.S. These tests revealed that Hall had indeed been poisoned—and Bessels became the prime suspect. But the arsenic found in his body may have been administered by Hall himself—at a time when arsenic served as both medicine and poison. Hall's body, and the mystery of his death, remain buried in northwest Greenland.

The failures of Kane, Hayes, and Hall, to say nothing of the U.S. Navy's disastrous *Jeannette* expedition north of Siberia under the leadership of George Washington De Long—along with many other Arctic disasters such as the German *Hansa* expedition—had finally led scientists to reconsider the value of such expeditions. A decade after Hall's death, in 1881, in an attempt to better understand the physical dynamics of the Arctic, several nations agreed to establish fixed meteorological stations in the high north to record synchronous weather observations across a full year.

The American contributions to this 'First International Polar Year' were stationed in two places, at Point Barrow in Alaska, and on the northern shores of Ellesmere Island in a complex effort led by a U.S. Army officer, Adolphus W. Greely. Greely was seventeen when he began a rapid climb from the enlisted ranks to a commission as a second lieutenant during the Civil War. He was commissioned into the regular army in 1867 and promoted lieutenant in 1873. By that time, a British-born officer in the U.S. Signals Corps named Henry W. Howgate had taken an interest in Arctic exploration and successfully lobbied for funding for a combined scientific and colonizing expedition to the shores of the Lincoln Sea.

Howgate's initial attempt to explore Greenland was a failure but did include several who would later join Greely. His second effort, in an unsuitable ship named *Gulnare*, did not sail with the blessing of the U.S. military and its leader, Lieutenant Gustavus A. Doane, later wrote sarcastically that it was the very first "acknowledged failure in Arctic annals. We did but little, but left a great many things undone requiring some moral courage to refrain from doing. We did not change the names of all the localities visited, as is customary, nor give them new latitudes to the bewilderment of the general reader. We do not dispute anyone's attained distance nor declare it impossible that he should have been where he was. We did not hunt up nameless islands and promontories to tag them with the surnames..." (New York Times, 1881).

When Howgate was accused of embezzlement by the army in 1880, planning for his third attempt at an Arctic colony, called the 'Lady Franklin Bay Expedition,' was already well underway in support of the International Polar Year. Howgate was arrested and charged just as Greely and his expedition sailed for the north (see, for example, Guttridge, 2000; Robinson, 2006).

Greely's crew was made up largely from Howgate's Signal Corps, soldiers typically responsible for weather reports and telegraph wires. The originator of the Signal Corps, Albert J. Myer, had also proposed a system of international weather stations and tried to integrate his ideas with those of the Austro-Hungarian polar explorer (and co-discoverer of Franz Josef Land), Karl Weyprecht. Both Weyprecht and Myer had died by the start of the Greely expedition and its association with the First International Polar Year, but it's programme of weather recordings, along with regular collection of data on air and sea temperatures, terrestrial magnetism, and the aurora, was one they had largely written and promoted at international conferences. Nowhere in these instructions were any hint of trying to reach the North Pole, which Weyprecht had concluded was a near-total waste of time (Fig. 5.5).

The expedition ship, a 190-foot-long barkentine named *Proteus*, arrived at Disko Island off western Greenland on 13 July 1881. There it was joined by a peripatetic American doctor of French extraction named Octave Pavy. For years Pavy had promoted dreams-*cum*-plans to reach the North Pole via an open polar sea and for as many years had been disappointed by the response of various scientific societies to his schemes. When Henry Howgate needed a doctor for his *Gulnare* expedition, Pavy saw his chance and joined the expedition, apparently in the wildly mistaken view that he would assume command of Howgate's planned Greenland colony.

When the *Gulnare* expedition failed, Pavy remained in Greenland, and when Greely arrived, Pavy was only too ready to join the new expedition as naturalist and doctor and to pronounce to one and all that only he possessed any Arctic experience whatsoever. Part of Greely's plans, to search for any potential traces of the wreckage of the *Jeannette*, would serve Pavy well as a cover for his true ambition to reach a new farthest north.

At Upernavik, the sealskin suits and dog teams and drivers Greely expected were in short supply. Two local hunters were rounded up for the expedition but neither spoke a word of English. So it was that this misaligned group—nineteen Signal Corps soldiers along with three civilians hastily mustered into the military, a French-American doctor, and two Greenland natives—made off for the northern shores of Ellesmere Island. Only Pavy and the Greenlanders had any experience with small boats, and the expedition's lone steam launch was looked after by an alcoholic sergeant.

Greely set up his encampment on Ellesmere Island on the shores of a bay named for Lady Jane Franklin, where George Nares had previously wintered. The site was just a few miles west of where Charles Francis Hall lay buried on the shores of

Fig. 5.5 Members of the Lady Franklin Bay Expedition prior to embarking in 1881. The image of Dr. Octave Pavy was added as he joined the expedition in Greenland. Three other portraits were also altered as the team was changed after the original portrait was taken (credit: Moses Parker Rice, photographer (public domain, via Wikimedia Commons and the National Oceanic and Atmospheric Administration)

Thank God Harbor in Greenland. Hall had by then been joined in the frozen ground by two members of the British Arctic Expedition of 1875-76 under Nares. Nares' efforts had reached a new farthest north record of 83°20′26″ but otherwise his expedition had largely been a failure (Fig. 5.6).

Greely constructed a large military-style barracks, named Fort Conger after a supportive U.S. Senator from Michigan, and commenced the scientific work scheduled to go on continuously until the retreat of the expedition two years later in the summer of 1883. In the spring of 1882, Greely departed from his orders to authorize Pavy to make a northern sledge journey to test whether the British mark under Nares could be bested. Pavy returned in disappointment, but with confirming that he had seen nothing like an 'open polar sea' north of Ellesmere Island.

Pavy's experience led Greely to shift his effort at a farthest north from Ellesmere to the north coast of Greenland. Accordingly, a week after Pavy's return, Lieutenant James Booth Lockwood, along with Sergeant David L. Brainard, set a new farthest north of 83°24′, just a few miles north of the British mark, at what was named Lockwood Island off the northern coast of Greenland. It marked the only notable geographic triumph of the expedition.

At this point, Greely had reason to be satisfied with his work as he awaited the ship that would provide relief to the expedition that summer. In Washington,

Fig. 5.6 East view of Fort Conger, Grinnell Land, 20 May 1883 (public domain via the National Oceanic and Atmospheric Administration)

however, the job of sailing to Greely's relief was given to a Civil War officer now reinstated as an army private, who left the capitol late and without any approved funding from Congress. By the time the funds were approved, half of the private's contingent had deserted, and he was forced to charter a substandard vessel in Newfoundland, the *Neptune*. The ship did not arrive off Disko until mid-July of 1882. Reaching the entrance to Smith Sound a week later, the Americans found a wall of ice more than a dozen feet thick. Before leaving the area for home, the *Neptune* party landed a small depot of emergency supplies at Cape Sabine on Pim Island off the east coast of Ellesmere Island on Nares Strait, fully 200 miles south of Greely's Fort Conger.

At Fort Conger, Greely by the end of August had given up on seeing a relief ship in 1882 and accordingly prepared his men for a second wintering. Greely even attempted forays in early 1883 to best his own expedition's farthest north, but these ended in disappointment on the thin spring ice off northern Greenland. There was little more to do but wait for the relief ship expected that summer.

What occurred in the summer of 1883 became a major embarrassment to the United States and a growing tragedy for Greely and his crew. The same *Proteus* that brought Greely to Lady Franklin Bay was chartered to bring him home. But the army officer sent to lead the expedition overruled the civilian captain of the *Proteus* and ordered the ship into a dangerous stream of ice at the mouth of Smith Sound before the ice had cleared. *Proteus* was promptly nipped and sunk. The crew scrambled to leave a few scattered supplies behind for Greely as they themselves frantically sought to escape south in small boats. A small navy vessel sent to escort *Proteus* was now forced to rescue the rescuers. Greely and his expedition were thus forced into a third winter in the Arctic, their morale falling as their desperation increased.

With promised relief ships unable to reach the men at Lady Franklin Bay for fully two years in a row, the expedition now descended into chaos, from embarrassment into national catastrophe. The U.S. Navy was called in to mount a four-vessel rescue of the army expedition but, by the time they reached Greely's retreating men at Cape Sabine, only six of the original twenty-five were still alive. Besides Greely himself, these included David Brainard, who had participated in the farthest north record north of Greenland.

The tragedy for Greely and his men might have been avoided by the selection of an experienced army veteran of Arctic travel like Frederick Schwatka, instead of an officer completely inexperienced in the High North like Greely. But Schwatka was soon to be dispatched on a survey of the Yukon River, eventually leading to another triumph for this remarkable officer when he successfully led a convoy of rafts from the headwaters of the river at Chilkoot Pass and then more than 1300 miles to the Bering Sea (Figs. 5.7 and 5.8).

The retreat of Greely occurred almost simultaneous with the arrival in the Arctic of a twenty-five-year-old German-American scientist by the name of Franz Boas. His studies of the Inuit of Baffin Island, chronicled in the monograph *The Central Eskimo* (1888), pioneered a new view of the Indigenous peoples of the Arctic. Boas saw the Inuit as every bit the equal of the explorers who for so long had relied on them for survival while denigrating virtually every aspect of their existence.

Figs. 5.7 and 5.8 The graves of Adolphus Greely and David Brainard at Arlington National Cemetery, September 17, 2025 (photos by the author)

One American who almost certainly read Boas was a twenty-six-year-old doctor from New York who had just lost his wife of 2 years. In his grief, he thereafter sought out the most remote parts of the planet. On an island off northwest Greenland, he was destined to have both his greatest triumph and, courtesy of the biggest sonofabitch in the history of American exploration, his most bitter failure. Along the way, Dr. Frederick Cook would transform American media for all time.

References

Bourke, J. G. (1891). *On the border with crook*. Charles Scribner's Sons.

Guttridge, L. F. (2000). *Ghosts of cape Sabine: The harrowing true story of the Greely expedition*. G.P. Putnam's Sons. See also: Robinson, Michael F. 2006. *The Coldest Crucible: Arctic Exploration and American Culture*. Chicago: The University of Chicago Press.

Loomis, C. C. (1971). *Weird and tragic shores: The story of Charles Francis hall, explorer*. Knopf. See also: Barr, ed. 2016. *Polaris: The Chief Scientist's Recollections of the American North Pole Expedition, 1871–73*. Calgary: University of Calgary Press.

New York Times. (1881). *Polar sarcasm: Lieut. Doane, of the Howgate expedition, presents an ironic report*. April 11, 1881.

Pierce, B. M. (1868). *A report on the resources of Iceland and Greenland*. U.S. State Department.

Chapter 6
Evelyn Briggs Baldwin, Robert Peary, Frederick Cook, and the Battle of Annoatok Island

Abstract The American obsession with the North Pole reached its climax between 1890 and 1910, with both base camps and relief camps pf U.S. polar expeditions established across Greenland. The final confrontation over priority at the North Pole came down to Robert Peary and Frederick Cook, and the continuing mystery of exactly what took place on a small island off the Greenland coast in the fall of 1909.

Keywords North Pole · Greenland · Greenlandic Inuit · American route · Annoatok · Cape York · Meteorite · Shannon Island · Ejnar Mikkelsen · Bass Rocks

> *Greenland is in fact but a Titanic mountain dam holding in check a mighty sea of ice. Where there are breaks in the dam, the ice pours through in a resistless glacier torrent and descends to the sea.*
>
> —Robert Peary, 1891.

The public and congressional outcries and investigations over the failures of Hall, De Long, and Greely would have far-reaching consequences for future American polar explorers. Not until the strategic value of the Arctic was rediscovered during World War II would any significant public money be spent on its exploration or exploitation. Until then, American explorers—beginning most notably with Robert Peary and then his many contemporaries and competitors—would have to rely on their own wits to secure private donations. Along with newspaper sponsorships, advertising tie-ins, and the backing of newly formed special interest organizations like the National Geographic Society, the Explorers Club, and the Peary Arctic Club, no explorer after Greely could seriously hope to sail north before he became adept at generating publicity and raising funds in this new and bewildering commercial universe (Fig. 6.1).

At the same time, young men of Peary's generation were enticed northwards by dramatic accounts of American polar exploration. In 1885, with future American polar explorers Peary, Frederick Cook, Walter Wellman, and Evelyn Briggs Baldwin

P. J. Capelotti, *Lost Greenland*, Springer Polar Sciences,
https://doi.org/10.1007/978-3-032-27841-8_6

Fig. 6.1 The *Polaris* entering the ice (from *The Marvellous Wonders of the Polar World* (1885))

all in their twenties, an author by the name of Herman Dieck edited *The Marvellous Wonders of the Polar World*, a compendium of first-hand expedition accounts, with 500+ pages on the Arctic and another 200 on the Antarctic, 'PROFUSELY ILLUSTRATED WITH SCENES AND INCIDENTS IN THE POLAR REGIONS AND PORTRAITS OF ARCTIC HEROES FROM ORIGINAL SKETCHES AND

PHOTOGRAPHS.' The illustration of Hall's *Polaris*, as one example among several, revealed the very portrait of a daring voyage to an otherworldly northern idyll of cathedral ice, clear starry skies, chattering gulls and basking seals.

Such dramatic illustrated volumes were all but guaranteed to entice young men to follow in the sledge tracks of Kane and Hayes, Hall and Greely, lack of government funding be damned. Perhaps the most astounding success in this arena was in fact a government meteorologist from Kansas by the name of Evelyn Briggs Baldwin. Baldwin would have been twenty-three when Dieck published his edited volume. Before he was killed by a passing automobile while crossing a street in Washington, D.C. in the fall of 1933, he made certain that his family monolith in Oswego Cemetery in Labette County, Kansas, was incised with the only two words he ever wanted attached to his name: "Arctic Explorer." It makes for an unusual sight, reached as it is via a dirt path that runs along wide, tree-rimmed field of southeastern Kansas (Fig. 6.2).

In his mid-twenties, Baldwin worked as a teacher and school administrator in Oswego before abruptly resigning, "having decided to devote more time to scientific research and travel." He went to work for the U.S. Government as an Assistant Observer for the U.S. Weather Bureau in Nashville, Tennessee. It was in Nashville in the early 1890s that Baldwin began to seriously study Arctic literature and, as he wrote, "to lay plans for work within the Arctic circle" (Capelotti, 2016: 209).

Baldwin's plans matured soon after, in 1893, when he used his experience as a meteorologist and his newfound knowledge of the literature of Arctic exploration to gain membership in Robert Peary's second expedition to northwest Greenland. Peary's first visit to Greenland, at the age of thirty, was just that, a brief penetration of the ice cap in 1886 financed with $500 of his mother's money and enabled by six month's leave from his duties as a civil engineer with the U.S. Navy.

It was 5 years before that Peary reached the ice again, this time a legitimate expeditionary attempt in 1891 to investigate whether Greenland was an island or a peninsula reaching to the North Pole. Peary was joined by his new wife Josephine, his servant and constant companion the African-American son of Maryland sharecroppers, Matthew Alexander Henson, and a small and *au fait* assortment of expedition members that included a young Norwegian ski expert named Eivind Astrup, and a recently widowed surgeon and ethnologist from New York, Dr. Frederick A. Cook. Peary broke his leg just as he reached northwest Greenland but went ahead with the expedition anyway. With Astrup, he reached the northeast corner of Greenland the following summer and claimed that he had proven Greenland to be an island. Even more rashly—and against Astrup's wise cautions—Peary decided that even more land lay to the north of Greenland, possibly including an icy road reaching all the way to the North Pole.

Tasting real fame for the first time after his announcement of the insularity of Greenland, Peary convinced the Navy to grant him 3 years leave for a new expedition. He intended to follow the small-scale 1891–1892 expedition with a much-larger effort to Greenland that would include his now-pregnant wife as well more

Fig. 6.2 The Baldwin family grave marker, Oswego Cemetery, Oswego, Kansas, in 2024 (photo by the author)

than a dozen others, eight of whom Peary planned to lead across the Greenland ice cap to try for the North Pole.

Peary needed a new man to handle weather recordings, since John Verhoeff, the meteorologist from his 1891–92 expedition, had disappeared without a trace in Greenland (see, for example, Robert M. Bryce, pp. 86–87). Verhoeff's family was hardly comforted by Peary's evasive answers as to the circumstances and his evident lack of enthusiasm for mounting a proper search for the man). Evelyn Briggs

Baldwin, dutifully recording the weather in Tennessee while he studied Arctic literature, became Verhoeff's replacement.

Frederick Cook had temporarily left Peary's orbit, but Eivind Astrup and Matthew Henson were both back and, on June 26th, 1893, the new expedition left Philadelphia and reached Inglefield Gulf in early August. There, on the shores of Bowdoin Bay, Peary's men constructed their winter quarters: Anniversary Lodge. It was here that Baldwin set up his self-registering meteorological instruments and kept notes on weather and aurora phenomena for an entire year, from August 3, 1893, to August 1, 1894 (Fig. 6.3).

The expedition was not a happy one. A large ice cap party, including Baldwin, left Anniversary Lodge on March 6, 1894, and ran directly into bone-rattling cold and skin-carving winds that stopped both men and dogs in their tracks. Many of the dogs went into a kind of Arctic mania called *piblockto* and were shot. "So intense was the suffering of the poor dog which first fell victim to the scourge, that he almost gnawed off his legs" (Capelotti, 2016: 210–211). Astrup contracted food poisoning from tainted pemmican and along with another expedition member had to be escorted back to the base camp. Two more men were forced back soon thereafter.

Only four men remained on the ice cap with Peary, and these included Baldwin and the remarkable Henson, a figure of titanic abilities in the Arctic and the only member of Peary's numerous polar attempts to master both dog-driving and the complex language of the indigenous Greenlanders living in and around Etah in northwest Greenland. They promptly trudged into a brutal spring storm, one that froze the dogs in their tracks.

> Baldwin's anemometer, barograph and thermograph, which, as a result of his ingenuity and perseverance, had kept on recording throughout the storm, showed that for thirty-four hours the average wind velocity had been over forty-eight miles per hour, and the average temperature about -50° Fahr., with a minimum of over -60° Fahr. When these figures are considered in connection with our elevation of some 5,000 feet, the unobstructed sweep of the wind, and the well-known fact that ice-cap temperatures accompanied by wind are much more trying to animal life than the same temperatures at sea level, it is believed that the judgment will be that this storm beat the record as the most severe ever experienced by any Arctic party (*ibid*: 211).

To go on with so few men and with more dogs dying of exposure and *piblockto* was pointless, so Peary turned the party around on April 10, 1894, just 128 miles from Bowdoin Bay. The men descended to the base camp covered in body lice and suffering from various degrees of exposure and snow blindness (Bryce, 1997: 132). Of ninety-two dogs that started the journey, only a shocking twenty-five remained and of those less than half were still able to do any work.

The quote above, purportedly from Peary, appears in Baldwin's fund-raising tract, *The Search for the North Pole*. It goes on with fulsome praise for Baldwin's "superior qualifications [and] depth of his character" as well as his "perserverance [sic], ingenuity, courage and skill, [and] power of endurance" (Capelotti, 2016: 210–211). Baldwin lists the source for this superlative quotation as "a recent college publication." It is more than likely that the author was Baldwin himself, as Baldwin

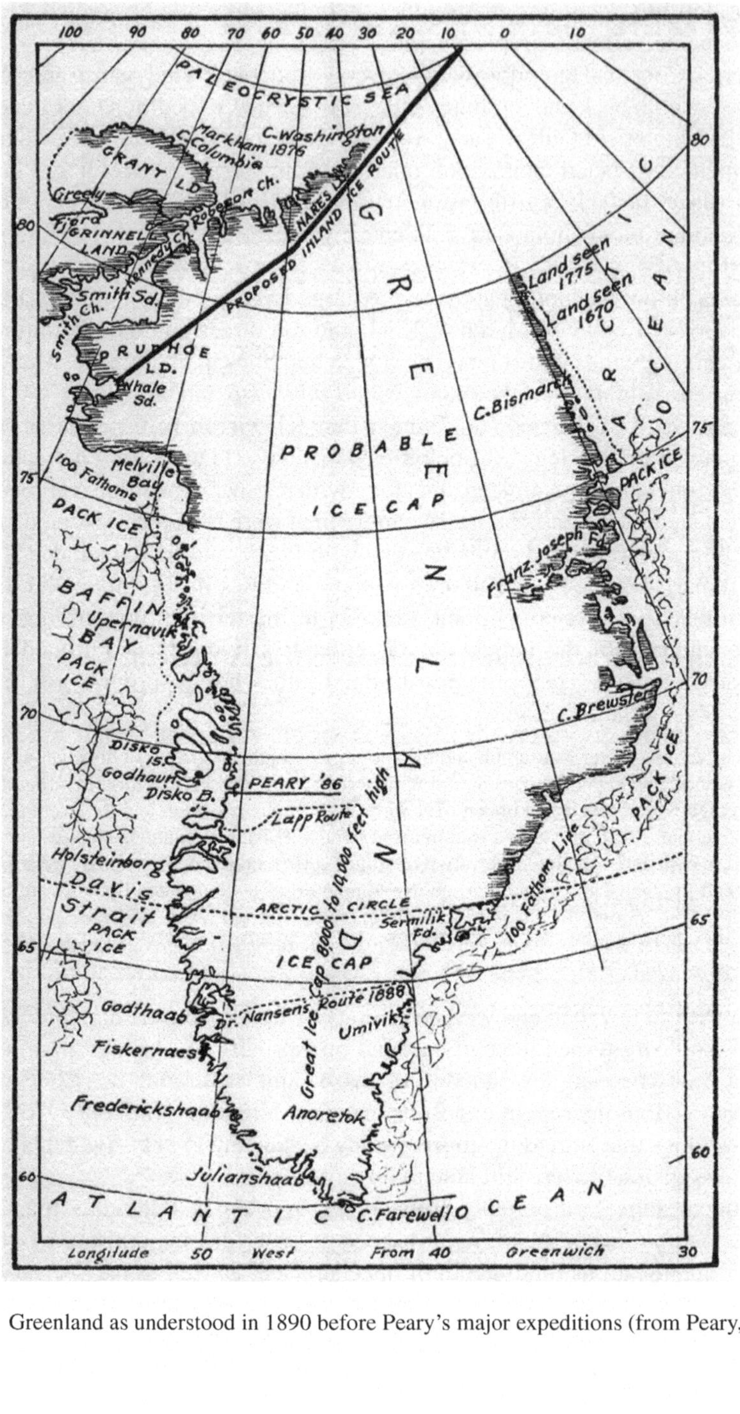

Fig. 6.3 Greenland as understood in 1890 before Peary's major expeditions (from Peary, 1891)

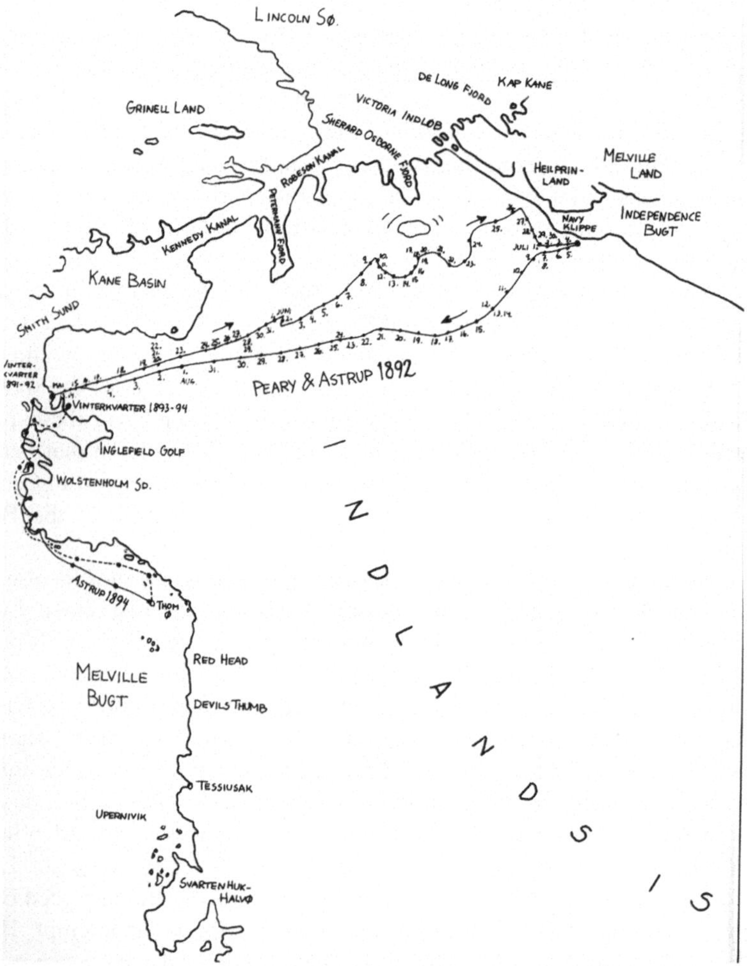

Fig. 6.4 Eivind Astrup's map over his travels over Northwest Greenland with Robert Peary in 1891/92 and to Melville Bay in April 1894 (Wikimedia commons)

had fully absorbed the new fine art of self-promotion in search of Arctic sponsors (Fig. 6.4).

In the curriculum vitae-*cum*-autobiography produced for one of those potential sponsors, a baking soda magnate named William Ziegler, Baldwin aimed a few passive-aggressive barbs at Peary for turning back after the great storm and hinted that things would have been different had he been in charge. Baldwin wrote of his disappointment at being ordered to retreat, calling it "the most bitter moment of my life. However, as Mr. Peary was in command, I do not wish this to be considered as criticism—only stated to show that we returned through no inability personally considered to proceed on the journey" (*ibid*: 211).

According to dual Peary/Cook biographer Robert Bryce, Peary thereafter retreated with his wife to their segregated living quarters at Anniversary Lodge, there to dine on a specially secreted cache of foods while the men in their freezing quarters subsisted on seal and walrus meat (Bryce, 1997: 133). It was a perverse lesson in expedition leadership that Baldwin would never forget and, almost from the moment he was given the chance to lead his own parties in the Arctic, he would reveal many of the same selfish and malevolent traits he had learned from his first Arctic mentor.

When Peary's relief ship *Falcon* arrived in early August, Peary sent most of his expedition home, including Baldwin and Eivind Astrup. Since 1894 had been such a disappointment, Peary and two others, including Matthew Henson, would stay on for another year. Baldwin concluded his year-long effort to record the weather at Bowdoin Bay and, along with Mrs. Peary, joined the remainder of the expedition sailing south. Back in Philadelphia, the men were shocked to learn that Peary had sent a letter ahead of the *Falcon* to announce that he was carrying on in Greenland despite being abandoned by his men. It was a shocking stab in the back that Baldwin never forgot.

To Eivind Astrup, Baldwin wrote in October of 1894 to ask if the Norwegian had seen Peary's defamatory letter. "You bet I did," the Norwegian responded. "Wonder if he will use the word "desert" about us, when he gets home next year. It probably would not pay him. Particularly if the most of us are out again next season, [thus] proving that we are just as good as him… My dear Baldwin, I hope you will show up your colors on the icefields next season, and I promise you that I shall show mine, the "dog drivers," and if we should not be successful next year, we are bound to be so one of the following 25 seasons" (Capelotti, 2016: 212).

Baldwin parlayed his year with Peary into a second-in-command billet with Walter Wellman's attempt to reach the North Pole from Franz Josef Land in 1898–1899. The expedition was a fiasco, not least because Wellman placed Baldwin in charge of the construction of an advance base for the polar attempt. Baldwin proceeded to advance less than a hundred miles north of the base camp, on the wrong side of a waterway, and then abandoned two of his Norwegian companions at the base for the winter—one of whom died a long, agonizing death from exposure.

Nevertheless, Baldwin turned even this disaster to his advantage, securing from William Ziegler an essentially unlimited budget to launch a new attempt on the North Pole in 1901.

Newspapers reported that Baldwin would venture north at the head of "the largest transport train in the history of Arctic exploration" (*ibid*: 238). By mid-April, an elephantine mass of supplies, "a huge mountain of hermetically sealed tins," carried atop dozens upon dozens of truck-loads—each hauled by four horses" (*ibid.*) began to arrive at the docks of New York for transshipment from the U.S. towards Norway.

Two of Baldwin's cousins were aeronautical experts who ran Baldwin Brothers Balloon Company in Quincy, Illinois. The brothers had made balloons for the World's Columbian Exposition in Chicago (*ibid*: 247) and with them Baldwin had placed a large order for small message-carrying balloons. On June 6, 1901, after posing for a photograph with Thomas S., Samuel, and Ivy Baldwin, near their giant balloon *Mars*, Evelyn Briggs Baldwin made his very first (and apparently the only) balloon ascent of his life (Fig. 6.5).

Emboldened by his first taste of aerial exploration, Baldwin finally let slip some of his plans to the Chicago newspapers. This was the first public notice that Baldwin planned to combine balloons with message buoys into a single balloon-buoy system. The balloons would not be used to carry the expedition northwards but to deliver expedition publicity *southwards*. The unmanned balloons would "mark the path of the expedition," though Baldwin failed to explain how he intended to set up new hydrogen-generating stations at each stopping point in the islands of Franz

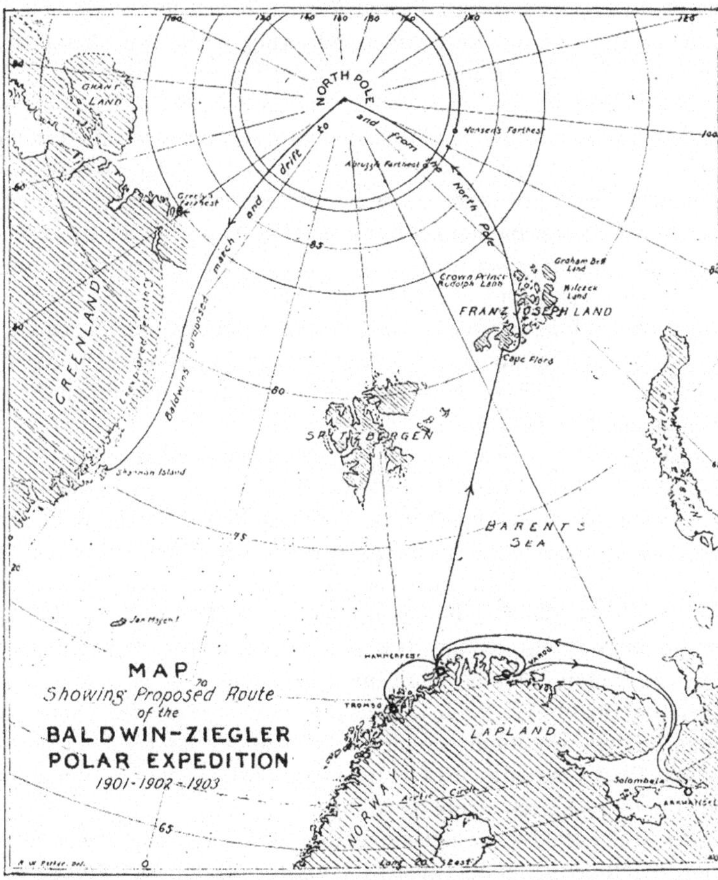

Fig. 6.5 Baldwin's proposed route to the North Pole and retreat via East Greenland. Balloon-inflation relief stations were laid at both Bass Rocks and Shannon Island (from: Evelyn Briggs Baldwin, 1901. "How I Hope to Reach the North Pole." The Windsor Magazine 15 (1): 59–69. London: Ward, Lock. December.)

Josef Land, much less while the expedition was on the move over the ice (Capelotti, 2016: 247–248).

On July sixth, Baldwin tasked Norwegian Johan Bryde, captain of the chartered vessel, *Belgica*, to land caches of supplies on the coast of northeast Greenland, in the event Baldwin returned in that direction after reaching the North Pole. In addition to the supplies, *Belgica* would carry prefabricated houses made by the firm of F.O. Peterson in Göteborg, the same company that built the prefabricated balloon house erected by the Andrée expedition at Virgo Harbor.

In directing where and how many of the expedition's supplies should be cached in Greenland, Baldwin ordered that two separate depot of provisions be landed on Koldewey and Shannon islands. At the first chosen site, Bryde was to erect the first two of the F.O. Peterson huts. These would store the expedition's perishable goods. A third hut would be placed at the second site.

The lists of the supplies carried by the *Belgica* reveal the scope of Baldwin's fantastic itinerary from the previous winter and spring, when he made repeated journeys around the U.S. and to Europe to make contacts and contracts with William Ziegler's money: dozens of boxes of Swedish conserves from Sveriges förenade konservfabriker AB in Lysekil—said to be the largest order a Swedish cannery had ever received (see: "Hedrande försvensk industry")—the Peterson expedition huts and over a hundred boxes of other sundries ordered from shipping agent Borlind, Bersén & Co in Göteborg, medicines from the UK, Army rations from Armour in Kansas City, Kansas, four boxes of flags from Astrup & Smith in Kristiania—all of this to go along with thirty tons of coal (*ibid.*: 251).

At the northernmost depot, Bryde was to leave behind two kayaks and a case of Kabo coffee; at the southernmost, two whaleboats and a packet of charts and some personal packages for Baldwin. Notices were to be left up and down the coast in both Norwegian and English (*ibid.*).

To top all this off, Bryde was to deposit at the Greenland stations several more hermetically sealed balloons from the Baldwin Brothers, along with the vats of acid and iron filings shipped from Dundee that were required to create hydrogen gas to inflate them. The acid and metal for the prospective balloon inflation operations in Greenland and Franz Josef Land had only been ordered in late June, along with instructions given to Baldwin on how to use the correct ratios of reactants to create the greatest amount of hydrogen gas. It was another example of an extremely complex operation being left to the very last minute (*ibid.*).

On the evening of July 16, *Belgica* shaped a course toward the islands of northeast Greenland. Bryde eventually created relief stations for Baldwin on Shannon Island and Bass Rocks in Northeast Greenland, almost too close to one another for either to be effective for a retreating expedition.

One recruit to the Baldwin-Ziegler expedition was a young cartographer with some previous Arctic experience, Ejnar Mikkelsen of Denmark. Mikkelsen, with a world of polar ambition, had previously tried to attach himself to the polar expeditions of Salomon Andrée, Otto Nordenskjöld, Otto Sverdrup, Carsten Borchgrevink, Adrien de Gerlache, and Eduard von Toll (Larsson, 2011: 153–154). Ultimately,

Baldwin's voyage to Franz Josef Land was a pathetic failure and the Greenland relief stations never used during the expedition. But Mikkelsen never forgot them.

Mikkelsen and University of Chicago geophysicist Ernest Leffingwell, who served as a surveyor for Baldwin, became fast friends and would later join forces to lead the 1907 Anglo-American Polar Expedition to Alaska's North Slope (see Mikkelsen 2005 and 1909). Mikkelsen then organized important geographic surveys of Northeast Greenland. He led an expedition to northeast Greenland to search for the bodies and journals of the 1906–08 *Danmark* expedition under Ludvig Mylius-Erichsen. In 1909, Mikkelsen found Mylius-Erichsen's journals but was himself trapped with a companion in northeast Greenland for three winters.

With appropriate irony, his desperate retreat from Greenland in 1912 was saved at the last moment by his knowledge of where Baldwin had ordered Johan Bryde to lay in large caches of supplies on Shannon Island and on Bass Rock in the spring of 1901. Taking refuge at the Baldwin hut on the southeastern point of Shannon Island in the winter of 1911–12, Mikkelsen and a companion were rescued by the Norwegian sealer *Sjøblomsten* the following July (Mikkelsen 1922: 136–139; Schledermann 1991: 352–354). In compiling his magnificent *North-East Greenland 1908–1960: The Trapper Era* (2008), the *Slædepatruljen Sirius* (Sirius Dog Patrol of North-East Greenland) veteran Peter Schmidt Mikkelsen visited and recorded the remains of both Baldwin relief stations. The octagonal huts at Bass Rock, photographed in 2004, and the hut at Kap Philip Broke, visited in 2007, were nearly intact after more than a century.

During his last visit to the Shannon Island site in April 1912, Mikkelsen found that animals had broken in "and foxes had dragged books out through the window and left them scattered all over the snow" (Mikkelsen 1922: 139). Like Leffingwell, Mikkelsen lived a long and productive life, passing away in his 90 s in Copenhagen in 1971. In 2009, the Danish navy named a Knud Rasmussen-class patrol vessel for him, the HDMS *Ejnar Mikkelsen*. Appropriately, the ship routinely cruises the waters around Greenland.

No American by himself, however, left behind more relics of his travels than Robert Peary, whose propensity for claiming everything in his path included at least two sons the result of a relationship he began in 1896 with a teenager named Ahlikahsingwah, the wife of one of his Greenlandic hunters, Peeahwahto. Their son Kali, born in 1906, lived long enough to journey in the 1980s to his father's homeland and visit both his father's prominent monument at Arlington National Cemetery, and his half-brother Robert Peary, Jr., in Maine. The visit was the result of research by the Harvard scientist S. Allen Counter, who had with immense patience prevailed on an extremely reluctant if not outright hostile Peary family to acknowledge that there existed in northwest Greenland a branch of their famous explorer's family that included some of the most well-regarded High Arctic hunters and sledge travelers (Counter, 1991).

Like Baldwin the government meteorologist, Peary, even though he was employed as a civil engineer for the U.S. Navy and carried no official rank, nevertheless carried on nearly all of his northern exploration at arm's length from any official government sanction. Heavily influenced by the writings of Kane and the

Swede Adolf Erik Nordenskiöld, Peary made his brief reconnaissance of the interior of Greenland in 1886. The three-week ascent to the Greenland ice cap ended after 125 miles with the realization of Peary's grossly inadequate training and equipment.

Before returning to his navy duties on a canal survey in Nicaragua, Peary met a Washinton, D.C. department store salesman by the name of Matthew Henson and together they formed a durable team that in 1909, with the essential aid of the Greenlanders Ootah, Egingwah, Ooqueah, and Seegloo, claimed the North Pole (for a full discussion of Peary's claim to the North Pole, see Herbert, 1989) (Figs. 6.6 and 6.7).

When the great Norwegian Fridtjof Nansen succeeded in 1888 where Peary had failed, in the first crossing of Greenland, Peary thereafter became deeply suspicious of anyone and any expedition that appeared in what he believed was his private domain to explore, claim, and name. He certainly came to consider northern Greenland as his and his alone to explore and exploit in the name of national honor, personal glory, and his own private profit.

Peary's 1894 effort to expand the scope of his previous expedition's geographic research failed, but it was during this time that Peary learned of the location of a meteorite at Cape York. This massive object had fallen from space thousands of

Fig. 6.6 Peary and Matthew Henson during Peary's survey work in Nicaragua in 1887, just before Peary's ambitions turned him north (from Hobbs 1936)

Fig. 6.7 Robert Peary, around 1912, after nearly two decades on Arctic ice (public domain via Wikimedia commons)

years earlier and fragmented as it spread across northwestern Greenland. It had been used by the indigenous Inughuit as a source of meteoritic iron for as long as anyone remembered.

Peary returned to Greenland in 1896 and 1897 to remove three meteorite fragments—the largest weighed over 30 tons—from the shores of Melville Bay and Meteorite Island. These he sold to the American Museum of Natural History in New York in order to finance more exploration. Along with the coerced removal of an Inughuit child, Minik, plus his father and five others, to New York for study by the museum staff, these events marked the disgraceful conclusion to half a century of polar ambitions along the 'American Route' to the North Pole first imagined by Elisha Kent Kane. Nearly all the Greenlanders were dead within a few months of arriving in New York, leaving Minik an orphan to wander between worlds until his death in the influenza pandemic of the fall of 1918 (Harper, 2000).

Peary left behind more than his progeny in northwest Greenland. In compiling his astounding *Expedition Relics from High Arctic Greenland* (2022), Peter Dawes, an emeritus research scientist at the Geological Survey of Denmark and Greenland in Copenhagen, identified a veritable archaeological horizon of material culture of just Peary's expeditions, one that could be studied apart from all other expeditions staged in the area. These included a surviving cache of coal cobbles near Etah and hypothesized to possibly derive from Peary's makeshift ship docking area. Dawes cleared up any lingering doubt when the cobbles were analyzed. It was not Tertiary coal of Greenland but Carboniferous coal from the Appalachian coalfields of eastern Pennsylvania.

The connections to the state where Peary was born continued with "Greenland's first and only railway coach…" (Dawes, 2022: 297). Peary brought to Greenland a caboose from the Erie Railroad to serve as a mobile headquarters. Sketches of Peary and his caboose were made by Russell Porter, who served on both the Baldwin and Fiala expeditions to Franz Josef Land, and made a visit to the Etah area as a member of a Peary relief expedition in 1899 (Porter, 1976: 58).

> In 1892 the Erie Railroad built a small four-wheel caboose numbered 4259 for use on short runs. The increase in freight tonnage made this style of caboose obsolete and after six years of service, following a crash that lifted it off its wheels, Erie Caboose #4259 was left in the Susquehanna, Pennsylvania railroad yard where it was used as a foreman's office in the sawmill section where ties were cut. While visiting a friend in that area, Captain Robert E. Peary saw the solidly constructed caboose and asked the Erie Railroad if he could use it on his upcoming voyage to the North Pole (https://www.shohola.com/railroad-caboose/).

Returned to the U.S. in 1902, the caboose, still minus its wheels, made a theme park attraction in Shohola, Pennsylvania, decorated with Arctic scenes, before being retired to Susquehanna, Pennsylvania, for display. It burned in a fire in 1940.

Dawes identified dozens of other traces of Peary's two decades in northwest Greenland, including artifacts from no less than seven Peary or Peary relief expeditions, and these do not include the tracks of his first ascent to the ice cap in 1886 or his claim to the North Pole itself in 1909.

Robert Peary died in 1920, and his body was carried through Washington, D.C. on a horse-drawn caisson to Arlington National Cemetery. When two years later the National Geographic Society sought a more prominent memorial, Peary was disinterred and his remains moved to one of the highest points short of Robert E. Lee's house itself at Arlington. The new memorial was an enormous globe with the North Pole prominently marked with a star.

His companion Matthew Henson, to the contrary, was buried in his wife's family plot at Woodlawn Cemetery in New York when he passed in 1955, forgotten by almost all including the same National Geographic Society that built so much of its reputation on his work. Once again, through the heroic efforts of S. Allen Counter, Henson and his American wife Lucy were removed in 1988 to Arlington, to rest adjacent to Peary and his monument. Neither monument or marker makes mention of the Greenland branches of both the Peary and Henson families (Fig. 6.8).

In death as in life, Peary had become the very symbol of American masculinity and achieved the fame he had sought all his life. His family made certain to memorialize that renown whenever and wherever possible. In 1932, his daughter and several former members of his expeditions journeyed to Cape York, just south of the modern Pituffik Space Base, and the area where Peary had used Greenlandic Inuit labor to drag their lone source of local iron tools away to his ship, to be sold to the

Fig. 6.8 Peary Monument and Henson grave at Arlington National Cemetery (photo by the author)

American Museum in New York. The memorial expedition constructed a 56'tall monument surmounted by an enormous 'P'. In shape it resembles the Washington Monument itself, and at exactly 10% of the height of that monument, would have perhaps seemed too short to satisfy the late explorer's famous ego (Peary-MacMillan Arctic Museum 2004).

Peary's fame can also be seen is his Pennsylvania hometown where, during the Great Depression, he was formally remembered by the place he had only lived in for 3 years. An elaborate 1937 dedication of Admiral Peary Park commemorated a project of President Franklin D. Roosevelt's Works Progress Administration, a monument administered today by the Pennsylvania Historical & Museum Commission (Fig. 6.9).

Of the traces of American expeditions in Greenland, however, none loom so large as the lost artifacts of Annoatok Island. It was to this wind-swept rock about 40 km north of Etah that Frederick Cook arrived in August of 1907. He had been offered transport to northwest Greenland by one John R. Bradley who, with his brother, had opened the Beach Club in Palm Beach, Florida, to cater to both men and—notorious for the time—women at his gaming tables. His resulting fortune from owning casinos allowed him to indulge himself in his real passion, big game hunting around the world. Bradley had also hired an assistant to come along, a German immigrant seaman named Rudolph Franke (Bryce: 294).

Fig. 6.9 Statue of Robert Peary at the Admiral Peary Park in Cresson, PA. Peary's physique manages to ripple even under his heavy Inughuit anorak (photo by the author)

For Cook, the free ride to Annoatok was a godsend. There he encountered Inughuit hunters and their families living in about a dozen sealskin tents and half that many stone igloo, shelter for what at the time was thought to be the most northerly inhabited spot on the globe. At forty-two and approaching the end of his active polar exploration days, Cook immediately saw his last great chance at immortality in the rich resources of the area. "The wealth in food and furs of this place fixed my determination on this spot as a base for the polar dash. We were standing at a point within seven hundred miles of the pole. The strongest force of men, the best teams of dogs, and an unlimited supply of food…" (quoted in *ibid.*: 303).

Cook convinced Franke to remain behind with him at Annoatok when Bradley sailed homewards, promising a chance at the biggest prize of all: the North Pole. Packing boxes formed the walls of a small hut, which was further enclosed by turf walls, and the two men settled in for an Arctic winter. By December, Cook was writing home that he had throughout the fall assembled one hundred dogs and 15 native dog drivers and hunters who, along with Franke, would depart and cross the ice to Ellesmere Island in January, there to start for the pole. Cook's plans even appeared in the *New York Times* in early October of 1907 (*ibid.*: 315; 322).

From this moment, the entire tale becomes a massive confusion of conflicting goals, egos, stories and, in the end, the conspicuous *absence* of artifacts related to Frederick Cook on Annoatok Island. When Peary learned of Cook's plans, he exploded, accusing the doctor of appropriating "the services of the Eskimo, whom I have trained in the methods of protracted serious sledge work…" (*ibid.*: 323).

Cook meanwhile had packed up his papers and letters, along with the furs and narwhal tusks hunted for profit to pay for the expedition and locked them inside his hut on Annoatok. He and Franke and a small group of Inughuit crossed over to Ellesmere Island at the end of February 1908. Finding the conditions along the northern edge of Ellesmere both difficult and game scarce, Cook stunned Franke by sending him back to Annoatok "to guard our station with its supplies" (*ibid.*: 328). In early May, Franke received a letter from Cook, sent with a returning group of Inughuit. He had continued with just himself and two young Greenlandic Inuit named Etukishuk and Ahwelah. Cook informed Franke that he expected to reach the pole and return by the end of May, and for Franke to gather all the materials from the hut and carry them south to Etah, from where Cook would begin his triumphal return to civilization.

Cook then vanished, and when an American polar explorer turned up in August and demanded the contents of the hut at Annoatok, it was not Frederick Cook but Robert Peary, returned to northwest Greenland to slay both the North Pole and Dr. Cook. Peary offered Franke paid passage back to the U.S. in exchange for all of Cook's possessions on the island, including his private letters and the key to his hut at Annoatok. Fearing a second winter in the high north, Franke had little choice but to agree. He regretted his decision almost immediately, as just a day after Peary sailed towards Ellesmere Island, a vessel called *Arctic* arrived in Etah carrying relief supplies and letters for Cook (*ibid.*: 333–334).

Before leaving for home, Franke gave the letters for Cook to a wealthy sportsman from Connecticut, Harry Whitney. Whitney planned to hunt game over the winter, and Franke told him of Cook's cabin on Annoatok. It was there, in the evening of April 15, 1909, that Frederick Cook and his two Greenlandic Inuit companions emerged from wherever they had been for 14 months. Whitney stood astonished, as by now it was assumed that Cook was long dead on the ice.

Soon, Cook entrusted his secret to Whitney:" I have been to the Pole" (*ibid.*: 340). Cook's story of his dash to the North Pole was complicated by his inexplicable decision not to return to Annoatok island immediately after the polar dash in 1908.

Instead, as Robert Bryce speculates, Cook had made no serious effort towards the pole but, realizing that he could not risk being seen by Inuit hunters so soon after his departure, he wandered south towards Lancaster Sound. Once there, enough time would have elapsed to make the polar dash plausible, and a passing Scottish whaler could carry him off to Europe to announce his triumph a full year before Peary. Instead, when Cook arrived and found no ship in sight, he was forced to overwinter at a desolate cave at Cape Sparbo on Devon Island, not to regain Annoatok until the following April (*ibid.*: 902–904). There he learned that Robert Peary was also on the scene, and in the middle of his own polar dash.

Unless he wanted to be pipped at the post, Cook knew he had to be first to get word of his claim to the world. He determined to reach the Danish settlements to the south and arrange transport to Copenhagen. Whitney, meanwhile, advised Cook to leave behind his heavy instruments and even his North Pole flag and polar sledge, as these would only slow him down. Cook agreed, leaving several boxes with Whitney and, on April 21, 1909, Cook hastened away.

Peary returned to land from his final attempt to reach the pole on April 23, and to his ship the *Roosevelt* soon after. It was, however, mid-August before the ship was freed from the ice and made its way slowly southwards towards Etah. There Peary picked up Harry Whitney for his return trip to the U.S., ordering the dismantling of Whitney's hut there and the distribution of all goods to the local Greenlandic Inuit. As for Cook's instruments and papers and his North Pole flag left on Annoatok, neither Whitney nor Peary makes any mention of them in their respective accounts (Whitney, 1910; Peary, 1910).

On Wednesday, September 1, 1909, as Roald Amundsen was in the middle of intensive plans for his own exploration of the north polar basin in the wake of Fridtjof Nansen's transpolar drift in the *Fram*, the morning newspapers arrived in the Norwegian capital with startling news. Frederick Cook, Amundsen's comrade from the 1897–99 *Belgica* expedition to Antarctica, had claimed to be the first human at the North Pole, in April of 1908. Cook's announcement shook the world, causing reverberations that last to this day. Even Cook did not realize the magnitude of his claim. He had sold the newspaper rights to his story to James Gordon Bennett's *New York Herald* for a mere $3000, a sum the Bennett would have been prepared to pay a hundred times more for.

More momentous news arrived 6 days later, on Monday, September 6. Robert Peary cabled from Battle Harbour, Labrador, that he had reached the North Pole in April 1909. Peary's allies in New York geographic circles immediately initiated a campaign to discredit Cook's claim. The rapid succession of claims and counter-claims was a catastrophe for none so much as Roald Amundsen. His expedition, only partially funded, was now less than 4 months from departure on a voyage few saw any further use for (Figs. 6.10 and 6.11).

At some point in the critical hours that followed, between the time he received the news of Cook, and the evening of Friday, September 10, after he placed an order for fifty Greenland dogs with the Danish inspector for North Greenland, Jens Daugaard-Jensen and effectively committed himself to Antarctica, Roald Amundsen turned from his long-planned exploration of the north polar basin to a rapid and

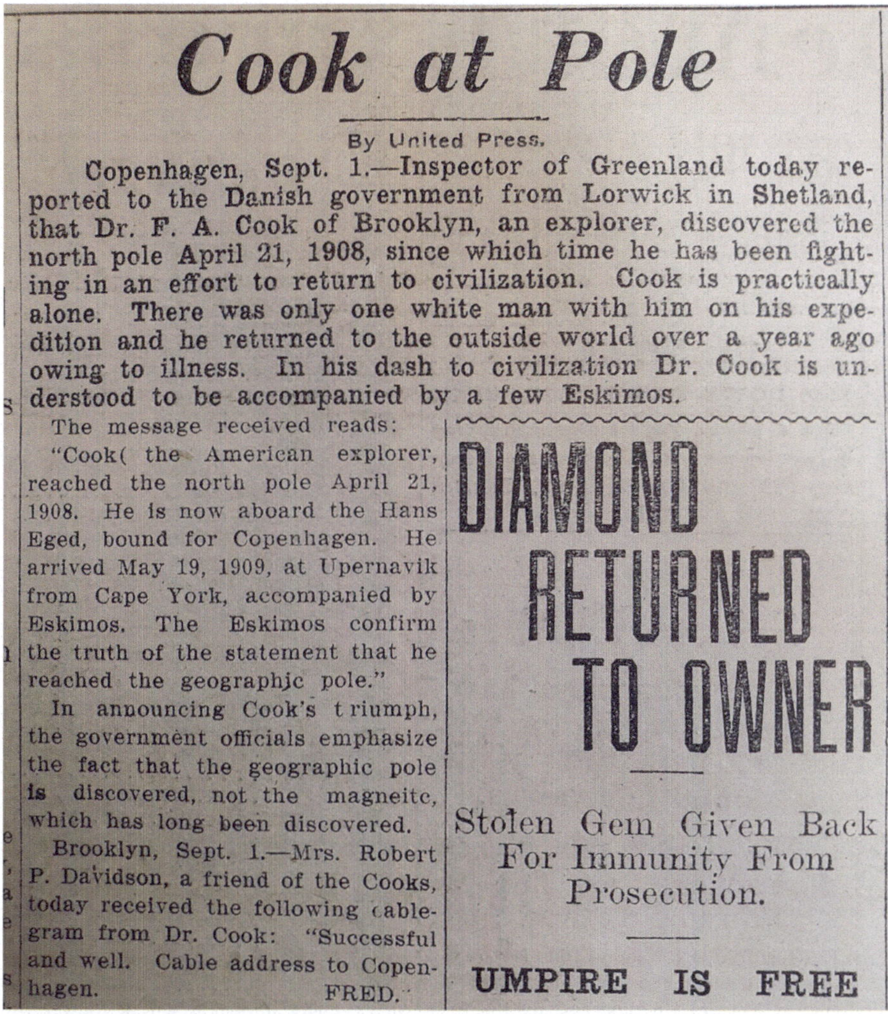

Cook at Pole

By United Press.

Copenhagen, Sept. 1.—Inspector of Greenland today reported to the Danish government from Lorwick in Shetland, that Dr. F. A. Cook of Brooklyn, an explorer, discovered the north pole April 21, 1908, since which time he has been fighting in an effort to return to civilization. Cook is practically alone. There was only one white man with him on his expedition and he returned to the outside world over a year ago owing to illness. In his dash to civilization Dr. Cook is understood to be accompanied by a few Eskimos.

The message received reads:

"Cook(the American explorer, reached the north pole April 21, 1908. He is now aboard the Hans Eged, bound for Copenhagen. He arrived May 19, 1909, at Upernavik from Cape York, accompanied by Eskimos. The Eskimos confirm the truth of the statement that he reached the geographjc pole."

In announcing Cook's t riumph, the government officials emphasize the fact that the geographic pole is discovered, not the magneitc, which has long been discovered.

Brooklyn, Sept. 1.—Mrs. Robert P. Davidson, a friend of the Cooks, today received the following cablegram from Dr. Cook: "Successful and well. Cable address to Copenhagen. FRED."

DIAMOND RETURNED TO OWNER

Stolen Gem Given Back For Immunity From Prosecution.

UMPIRE IS FREE

Fig. 6.10 September first, 1909, the start of a week Cook and Peary sent shockwaves through the media and caused Roald Amundsen to secretly change his planned North Polar Basin expedition to an attempt on the South Pole instead, placing him squarely in competition with Robert Falcon Scott of Great Britain (author's collection)

risky expedition to be the first human to stand at the geographic South Pole (Capelotti & Barr, 2013) (Figs. 6.12 and 6.13).

As for Peary, he had no more proof that Cook was a fraud than Cook had proof that he had actually reached the pole. Peary's own account showed that he had turned back the last competent navigator when he was still over 130 nautical miles from the pole and then managed to cover those remaining miles in barely 4 days. It all added up to perhaps the first time in American history when the news media

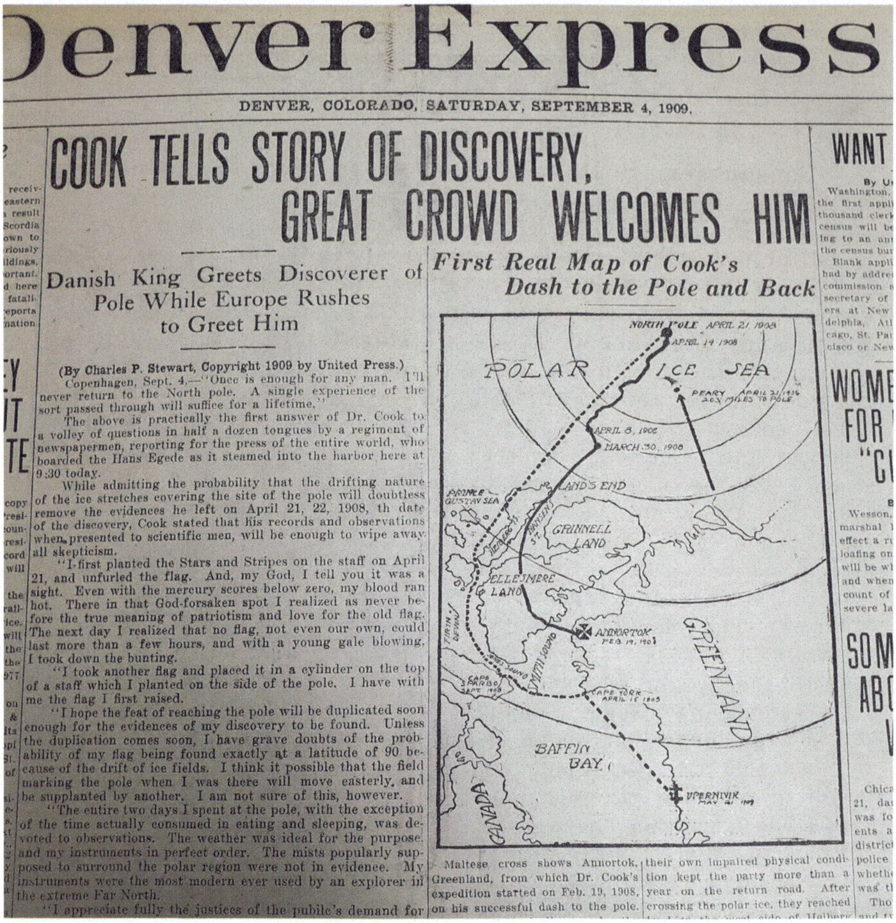

Denver Express

DENVER, COLORADO, SATURDAY, SEPTEMBER 4, 1909.

COOK TELLS STORY OF DISCOVERY, GREAT CROWD WELCOMES HIM

Danish King Greets Discoverer of Pole While Europe Rushes to Greet Him

(By Charles P. Stewart, Copyright 1909 by United Press.)

Copenhagen, Sept. 4.—"Once is enough for any man. I'll never return to the North pole. A single experience of the sort passed through will suffice for a lifetime."

The above is practically the first answer of Dr. Cook to a volley of questions in half a dozen tongues by a regiment of newspapermen, reporting for the press of the entire world, who boarded the Hans Egede as it steamed into the harbor here at 9:30 today.

While admitting the probability that the drifting nature of the ice stretches covering the site of the pole will doubtless remove the evidences he left on April 21, 22, 1908, th date of the discovery, Cook stated that his records and observations when presented to scientific men, will be enough to wipe away all skepticism.

"I first planted the Stars and Stripes on the staff on April 21, and unfurled the flag. And, my God, I tell you it was a sight. Even with the mercury scores below zero, my blood ran hot. There in that God-forsaken spot I realized as never before the true meaning of patriotism and love for the old flag. The next day I realized that no flag, not even our own, could last more than a few hours, and with a young gale blowing, I took down the bunting.

"I took another flag and placed it in a cylinder on the top of a staff which I planted on the side of the pole. I have with me the flag I first raised.

"I hope the feat of reaching the pole will be duplicated soon enough for the evidences of my discovery to be found. Unless the duplication comes soon, I have grave doubts of the probability of my flag being found exactly at a latitude of 90 because of the drift of ice fields. I think it possible that the field marking the pole when I was there will move easterly, and be supplanted by another. I am not sure of this, however.

"The entire two days I spent at the pole, with the exception of the time actually consumed in eating and sleeping, was devoted to observations. The weather was ideal for the purpose, and my instruments in perfect order. The mists popularly supposed to surround the polar region were not in evidence. My instruments were the most modern ever used by an explorer in the extreme Far North.

"I appreciate fully the justices of the pubilc's demand for

First Real Map of Cook's Dash to the Pole and Back

Maltese cross shows Annortok, Greenland, from which Dr. Cook's expedition started on Feb. 19, 1908, on his successful dash to the pole.

their own impaired physical condition kept the party more than a year on the return road. After crossing the polar ice, they reached

Fig. 6.11 Chart of Cook's claimed journey from Annoatok in northwest Greenland to the North Pole, with an extended retreat via Cape Sparbo. Cook replied to reporters that he hopes someone will duplicate his feat soon, otherwise the flag he left in a cylinder at the pole will be forever lost to the drift of the ice. Asked for actual proof, Cook demurs, stating that once his records were examined by "scientific men," his claim will prevail. When a committee of such men at the University of Copenhagen examined Cook's materials, they found them entirely lacking in navigational data and contained no proof that he had been at the pole. Peary was never required to submit his records to anyone other than one of his sponsors, the National Geographic Society, which proclaimed itself entirely satisfied and buried the evidence for 80 years (author's collection)

would chase a sensational story not in an honest attempt to get at the truth, but to exploit one side or the other of the drama of conflicting claims to sell more papers, magazines, books, and lecture tours. It opened the door to critics of the press to dismiss any report they found inconvenient, and reward editors who might be persuaded to publish stories with a predetermined slant.

ver Express

THE WEATHER
Denver and Vicinity—Fair tonight and Sunday; cooler Sunday.

DENVER, COLORADO, SATURDAY, SEPTEMBER 11, 1909. 1 CENT A COPY

THE RIVALS

"AN IMPOSTER" PEARY'S CHARGE AGAINST COOK

Experts on Arctic Affairs Declare Peary's Account of His Trip Is Convincing

By United Press.

New York, Sept. 11.—Peary, at Battle Harbor, sent a wireless message to the United Press containing a reiteration of his charge that Cook was never near the pole.

He asserts that at the proper time he will bring proof that the Brooklyn physician is an imposter.

He does not expect to arrive at Sydney before Sept. 15.

The United Press message reads: "Battle Harbor, via Marconi Wireless to Cape Ray, Sept. 10.—Accept thanks for your kind telegram of congratulation. Replying to your inquiry, I will say I found no trace of Cook at the pole. As a matter of fact, Cook hasn't been at the pole on April 21, 1908, or at any other time. This statement is made advisedly and will be backed by proof at the proper time.

(Signed) "PEARY."

In a supplementary message to the United Press of the same date, Peary says: ."See my Washington cablegram to white house, state department, navy department, and coast survey; also telegrams to Pres. Schurman, Cornell university.

(Signed) "PEARY."

The New York Times, under contract to publish Peary's narratives, today publishes a cable from Peary declaring that as soon as Cook makes his claims before a reputable geographical, scientific society Peary will be "in position to furnish material which may prove distinctly interesting reading for the public."

(By Herbert Bridgman, Secy. of the Peary Arctic Club, Staff Corre- | was reached in five dashes after he

Milkmen Fighting

Fig. 6.12 The *Roosevelt* reached the telegraph station at Battle Harbour, Labrador, in early September, from which point Peary and his allies launched on comprehensive assault on Cook's credibility (author's collection)

It could even be argued that the first week of September of 1909 was the start of much of modern journalism—sensational, profusely illustrated with drawings, photographs, and maps, and in many respects entirely partisan—had its genesis in the lost cache on the small, windswept island of Annoatok. As a result, journalists frantically charted boats to interview Peary in Labrador while maps of remote hunting camps like Annoatok of the Greenlandic Inuit suddenly appeared on the front pages of national and international newspapers.

Cook, for his part, would later claim Peary had refused to embark the instruments and papers he had placed in Whitney's care on Annoatok. And this after thieving all of the skins and furs Cook had secured to defray the cost of his expedition. Whitney, for his part, was now shoved to the center of the controversy. He telegraphed Cook on September 26 that "Peary would allow nothing belonging to you on board." It had all been moved to a cache near Etah (Bryce: 402). Cook told reporters that the records now presumably at Etah were originals, but that he gave copies to the University of Copenhagen, and that he would soon return to Greenland to retrieve them, but not just yet as it was too late in the season.

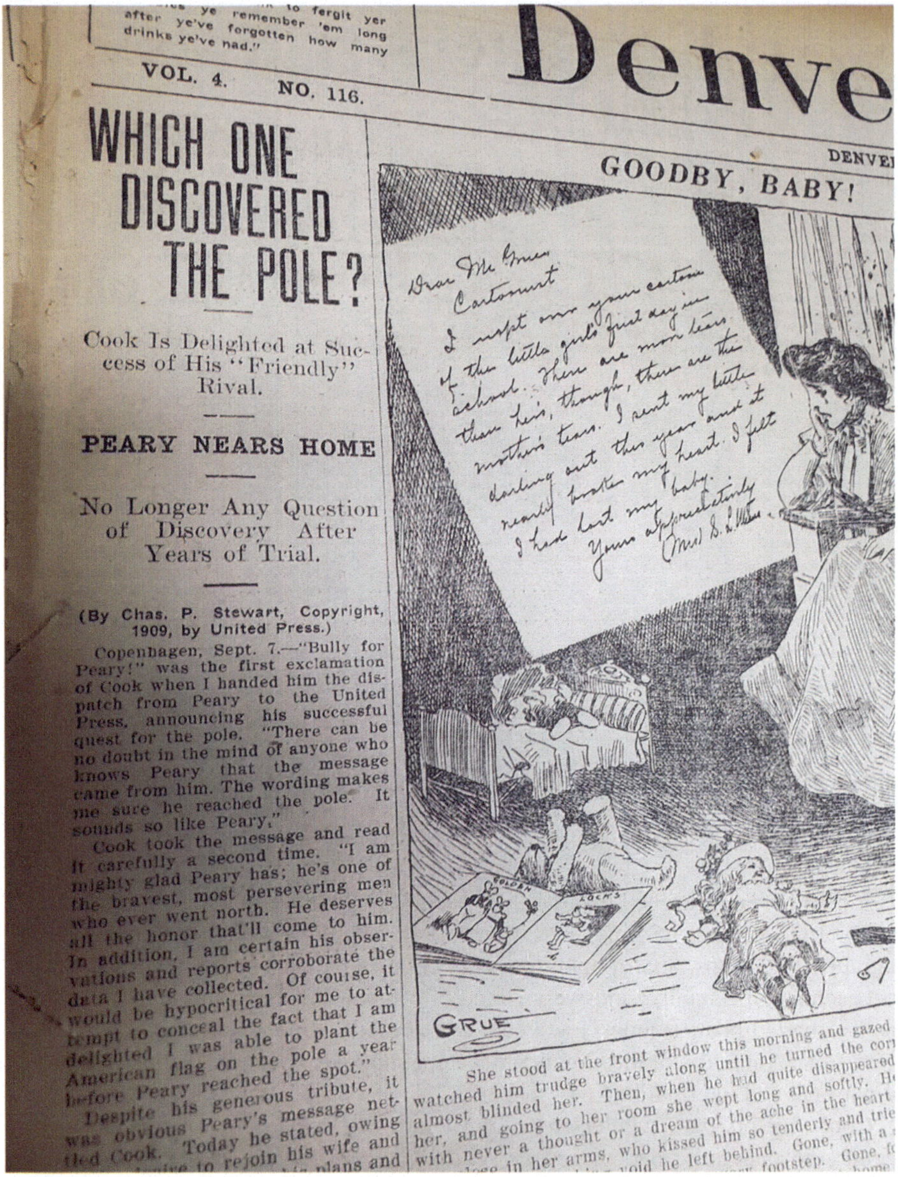

ye remember 'em long after ye've forgotten how many drinks ye've had."

VOL. 4. NO. 116.

Denve

WHICH ONE DISCOVERED THE POLE?

Cook Is Delighted at Success of His "Friendly" Rival.

PEARY NEARS HOME

No Longer Any Question of Discovery After Years of Trial.

(By Chas. P. Stewart, Copyright, 1909, by United Press.)

Copenhagen, Sept. 7.—"Bully for Peary!" was the first exclamation of Cook when I handed him the dispatch from Peary to the United Press, announcing his successful quest for the pole. "There can be no doubt in the mind of anyone who knows Peary that the message came from him. The wording makes me sure he reached the pole. It sounds so like Peary."

Cook took the message and read it carefully a second time. "I am mighty glad Peary has; he's one of the bravest, most persevering men who ever went north. He deserves all the honor that'll come to him. In addition, I am certain his observations and reports corroborate the data I have collected. Of course, it would be hypocritical for me to attempt to conceal the fact that I am delighted I was able to plant the American flag on the pole a year before Peary reached the spot."

Despite his generous tribute, it was obvious Peary's message nettled Cook. Today he stated, owing to ... to rejoin his wife and ... plans and

DENVE

GOODBY, BABY!

She stood at the front window this morning and gazed watched him trudge bravely along until he had quite disappeared almost blinded her. Then, when he had quite disappeared her, and going to her room she wept long and softly. H with never a thought or a dream of the ache in the heart ... in her arms, who kissed him so tenderly and trie ... void he left behind. Gone, with a ... footstep. Gone, f

Fig. 6.13 From Cook gets the Pole, to Peary claims the Pole, to Peary accuses Cook of fraud, to a lingering open question favoring Peary, all was set in motion in the first week of September 1909 (author's collection)

At first, none of it mattered. A standing room only audience greeted Cook at Carnegie Hall on September 27. Ten thousand awaited his train arrival in Philadelphia 2 days later. When Peary arrived in the U.S., he dismissed the Annoatok controversy immediately. It was, as Robert Bryce summarized Peary's remarks, "a scheme to either lose or destroy them, since he could not conceive how any man could have separated himself from such priceless possessions" (*ibid.*: 417).

Yet Annoatok Island would follow Harry Whitney to the end of his life. More than half of his obituary in the *New York Times* was spent retelling the tale of Cook's return from the pole to spend "two days at his base, Annotok (sic), near Etah. He left instruments, clothing and other effects with Mr. Whitney and went on to Upernavik, where he expected a boat" (*New York Times*, May 21, 1936). Whitney returned to the area at least two more summers to hunt. "Each time there were reports that he would seek the cache of Dr. Cook's instruments—which supposedly contained records that Mr. Whitney said he had never seen" (*ibid*).

In the end, whatever the truth of it, none of Cook's instruments or papers were ever found, if in fact they had been cached in northern Greenland. Like the cylinder containing an American flag he claimed to have left at the North Pole, the instruments and papers left on Annoatok Island became convenient excuses for Cook's inability to prove his claim while simultaneously placing the blame on Peary's malfeasance. And since that claim was never to hold up under scrutiny, the best thing Robert Peary could have done for himself was to welcome Harry Whitney aboard the *Roosevelt* while sealing all of Cook's materials in a secure room on the ship and leaving them untouched.

Peary fared better than Cook, but only because he was never subject to the intense scrutiny that fell on Cook throughout the fall of 1909. It would be more than 60 years after his death that Peary's own journals would be examined and found wanting. Had Peary gotten closer to the pole? Of that there was no question. Had he actually reached the spot? No one would ever be able to claim that with certainty, and for that Peary had only himself to blame. Roald Amundsen took careful note, and when he reached the South Pole 2 years later, he made certain his navigation was as perfect, and the spot itself marked as comprehensively, as he could possibly define it. His claim was also seconded, albeit reluctantly, when Robert Falcon Scott of Great Britain reached the exact spot a month later, and found Amundsen's tent—surrounded by black flags and surmounted by a Norwegian flag—waiting for him.

When the University of Copenhagen issued its damning report on Cook in December of 1909 Knud Rasmussen, who had sat in on the committee's work, declared that Cook had submitted "a most childish attempt at cheating" (Bryce 471). Cook's earlier claim to the first ascent of Mount McKinley was then similarly examined and also found wanting. The crowning disgrace would be a 1923 conviction on fourteen counts of stock fraud, which carried a prison sentence of nearly 15 years. Cook served 5 years and just prior to his death in 1940, received a pardon from President Franklin D. Roosevelt (Figs. 6.14 and 6.15).

Through it all, his two adoring daughters stayed by him through thick and thin, and his ashes were eventually interred between theirs in a columbarium at Forest Lawn Cemetery in Buffalo, New York. A large half-globe memorial sculpture titled

Fig. 6.14 Cartoonists, always a step ahead of pundits, concluded the dramatic first week of September 1909, with the expected result of the U.S. claim to the North Pole: land grabs, indigenous evictions, resource exploitation, and John D. Rockefeller immediately cornering the whale market and raising the price of oil (author's collection)

Fig. 6.15 Globe at Forest Lawn Cemetery in Buffalo, New York, commemorating Frederick Cook as the discoverer of the North Pole (photo by the author)

"Top of the World" was added outside the building, becoming a staple of cemetery tours, and offering no doubt as to who had first reached the North Pole from Greenland.

References

Baldwin, E. B. (1901). How I Hope to reach the north pole. *The Windsor Magazine, 15*(1), 59–69. London: Ward, Lock. December.

Bryce, R. M. (1997). *Cook and Peary: The polar controversy, resolved*. Stackpole Books.

Capelotti, P. J. (2016). *The greatest show in the Arctic: The American exploration of Franz Josef Land*. University of Oklahoma Press.

Capelotti, P. J., & Barr, S. (2013). Oppegård kirke, Oppegård, Norway: Fixing the location where Roald Amundsen revealed his 'minor diversion.'. *Polar Record* , on-line at. https://doi.org/10.1017/S0032247413000181

Counter, S. A. (1991). *North pole legacy: Black, White, and Eskimo*. University of Massachusetts Press.

Dawes, P. R. (2022). *Expedition relics from High Arctic Greenland*. Museum Tusculanum Press.

Dieck, H. (1885). *The Marvellous wonders of the polar world*. The National Publishing Company.

Harper, K. (2000). *Give me my father's body: The life of Minik, the New York Eskimo*. Steerforth Press.

Herbert, S. W. (1989). *The noose of laurels*. Atheneum.

Hobbs, W. H. (1936). Peary. New York: MacMillan.

Larsson, A. (2011). *...skönt att än en gang vara bland isen*. B4 Press.

Mikkelsen, E. (1922). Alabama-Expeditionen til Grønlands Nordøstkyst 1909–1912. København: Meddelelser om Grønland BLII.

New York Times. (1936). *Harry Whitney, 62, Explorer, is dead.* May 21, 1936, p. 23.

Peary, R. E. (1891). A proposed exploration of Northern Greenland. *Journal of the American Geographic Society, 23*, 161.

Peary, R. (1910). *The north pole*. Frederick A. Stokes.

Porter, R. W. (1976). The Arctic diary of Russell Williams Porter. In R. Herman (Ed.), *Friis*. University of Virginia Press.

Peary-MacMillan Arctic Museum. 2004. Building the Peary Monument. Retrieved December 31, 2025, from https://www.bowdoin.edu/arctic-museum/exhibits/2004/peary-monument.html.

Schledermann, P. (1991). Einar Mikkelsen (1880–1971). ARCTIC. 44. https://doi.org/10.14430/arctic1558.

Whitney, H. (1910). *Hunting with the Eskimos*. The Century Company.

Chapter 7
"Fish" Hassell, Parker "Shorty" Cramer, Charles and Anne Lindbergh, and the Search for a Northern Air Route to Europe

Abstract As Arctic explorers turned from sledges to aircraft in the 1920s, Greenland became the epicenter of the American ambition to establish a northern air route to Europe. Aeronautical expeditions that sought to pioneer this route met with both triumph and tragedy throughout the interwar period.

Keywords Greenland · Greenlandic Inuit · Copenhagen · Danish West Indies · U.S. Virgin Islands · 1917 Convention · U.S. Navy · Søndre Strømfjord · Port Burwell · Labrador · Davis Strait · Angmagssalik · Tasiilaq · Lauge Koch · Bluie West 8 · Kangerlussuaq

The fallout from the Cook-Peary feud in the fall of 1909 continued to color American interest in Greenland for decades. Barely one year following the infamous first week of September 1909, the American minister to Denmark, an Irish Catholic scholar named Maurice Egan appointed to the Copenhagen post by Theodore Roosevelt, sent to the Assistant Secretary of State what he called "a very important and very audacious suggestion" (U.S. Department of State Archives, 1910). In it, Egan argued for a radical solution to the growing border tensions around the globe: a land swap between Denmark, the United States, Germany, and the U.S. colonial occupation of the Philippines. Denmark would trade Greenland to the U.S. in exchange for the southern islands of the Philippines, including Palawan and Mindanao, which Denmark would then hand over to Germany in exchange for Germany returning to Danish sovereignty a part of the province of Schleswig.

> "It would mean, if carried into effect, the acquisition on our part of both Greenland and the Danish West Indies. The West Indies do not, as you will observe, appear in the propositions which I have gathered for you, but they might be made to appear. Greenland is, as you know, a Danish monopoly. It has never been exploited, although the Norwegians are clever enough to see its possibilities, as they already see what might be done with a lesser opportunity in Iceland.
>
> Greenland is looked on by the Danes very much as our people formerly looked on Alaska. The Government here is so much occupied with internal economic and political differences

P. J. Capelotti, *Lost Greenland*, Springer Polar Sciences,
https://doi.org/10.1007/978-3-032-27841-8_7

> in Denmark that it gives very little attention to the development of the resources of Greenland, which is practically *terra incognita* (*ibid.*).

Egan was well aware that such a potential swap was little more than an audacious land grab.

> "The possession of Denmark in Greenland to be ceded to the United States would exceed more than 30 times the area of the islands of the southern group of the Philippines, that is to say, that Denmark would in exchange for the above-mentioned part of the province of Schleswig, give territory in Greenland of over 300 times larger an area.
>
> Or, in other words, Denmark would, in exchange for a part of Schleswig, only 2,400 English square miles in area, give to America territory in Greenland as large as about one-fourth of the total area of the United States of America, or a territory as large as the whole of Mexico, or about 2½ times the size of Alaska (*ibid.*).

Egan's proposal was not acted upon, but he persisted in his efforts to combine the acquisition of Greenland with the U.S. desire to forestall Germany in the Caribbean by gaining the Danish West Indies for an American submarine port. The acquisition of the Danish West Indies had been mooted about the Department of State since 1899, when Seward's 1867 maneuvers in the Caribbean were revived. Discrete inquires extracted from the Danes the notion that perhaps four or five million dollars would suffice for the U.S. to purchase the islands. Negotiations stalled when the U.S. would not guarantee the right to U.S. citizenship to the islanders (U.S. Department of State Archives, 1901).

Maurice Egan's efforts were finally successful in 1917, when Denmark finally agreed to sell the islands to the United States for $25 million dollars, whereupon they assumed the name of the U.S. Virgin Islands. President Woodrow Wilson, however, was completely uninterested in placing Greenland into the bargain, arguing that "it's not worth anything at all. It's just ice. I don't want it" (ABC News, 2025). So, in the same treaty, a separate statement was appended from then-Secretary of State Robert Lansing declared: "In proceeding this day to the signature of the Convention respecting the cession of the Danish West-Indian Islands to the United States of America, the undersigned Secretary of State of the United States of America, duly authorized by his Government, has the honor to declare that the Government of the United States of America will not object to the Danish Government extending their political and economic interests to the whole of Greenland" (Convention between the U.S. and Denmark, 1917: 39 Stat. 170) (Figs. 7.1 and 7.2).

Robert Peary was among those entirely unsatisfied with Wilson's disinterest to obtain Greenland. Since claiming the North Pole, the newly commissioned Captain in the U.S. Navy began a decade of advocacy that combined the acquisition of Greenland with the growing threat of air power. The only way to successfully defend the American homeland, Peary argued, was the placement of U.S. air and naval bases in Greenland (quoted in Bartlett, 1940).

Peary's lieutenants from his many forays were also following in his wake. Peary's 1906 sighting of what he thought was a landmass out in the Arctic Ocean—which he named Crocker Land after a patron—was followed up by Donald MacMillan in

Fig. 7.1 As the Danish flag is lowered, the final Danish military parade gathers at Saint Croix, March 31, 1917 (source: public domain via Wikimedia Commons)

1913. Through one misfortune after another—including the alleged murder of one of the Inuit guides by American Fitzhugh Green—the expedition succeeded in identifying 'Crocker Land' as a Fata Morgana, an Arctic mirage (see: Welky, 2017). Both Green and MacMillan were sure it was land, but their Greenlandic Inuit were just as dismissive: it was nothing more than mist on the horizon (Hobbs 1936: 296–297). The expedition was not finally rescued until the arrival of Robert Bartlett in 1917.

In late 1918, at the conclusion of the First World War, Bartlett, another veteran of Peary's 1909 North Pole expedition, was tapped by now Rear Admiral Peary to lead

Fig. 7.2 A 1917 illustration for the Danish satirical publication *Klods-Hans*, by Alfred Schmidt, caricatures Woodrow Wilson as Uncle Sam 'adopting' the three islands of St. Thomas, St. John, and St. Croix, depicted as racial slurs of Afro-Caribbean children (source: Wikimedia Commons)

a series of polar flights from a proposed airbase at Etah. Peary would make the journey as base commander at Etah, while Bartlett led aircraft large and small as they surveyed unknown tracts of the Arctic Ocean all the way to the North Pole (New York Times, 1918). Again, Peary would proceed without any official government support, and sought instead to raise the necessary funds through the Aero Club of America, created in 1905 to promote aviation in the U.S. It was this same club that in 1919 offered the $25,000 Orteig Prize for the first aviator to cross the Atlantic in one flight: the prize claimed by Charles Lindbergh just eight years later.

The great aerial survey under Peary never came off. Peary died in 1920, but his dream of aeronautical exploration and national defense from a Greenland was picked up by General William E. 'Billy' Mitchell, who argued for bases in both Greenland and Iceland, and for an aerial Arctic expedition by the military airship U.S.S. Shenandoah, which was eventually canceled when the airship was damaged in a storm in early 1924 (Emmerson, 2011: 123). A half-hearted U.S. naval aviation detachment equipped with two-seater, open-cockpit biplanes and led by a Lieutenant-Commander by the name of Richard E. Byrd was attached to Donald MacMillan's 1925 Greenland expedition to Etah. Byrd's chief pilot was a U.S. Navy enlisted man, Floyd Bennett. MacMillan had little faith in the aircraft, and in fact within weeks all three of the biplanes were either wrecked or out of service in the harsh conditions. Byrd had fantasized about making an aerial dash for the North Pole, but in the end the aircraft could not manage even an advance refueling stop a few miles away on Ellesmere Island (Grover, 2001).

By 1929, the 1919 flight of the R-34 dirigible was still the only successful round-trip flight between Europe and the United States. Aircraft, still in their relative infancy, were limited by their short range to making short hops, from New York to Boston, Boston to Nova Scotia, Nova Scotia to Greenland, and so on, across to Europe. New York was the favored terminus for these sea hops, as it had been for Lindbergh, when Lindbergh flew the Atlantic solo in 1927. But Lindbergh carried no payload, and other aviators now sought to use similar technology on a different route to establish a regular, passenger-carrying airline across the North Atlantic.

Lindbergh's New York to Paris flight had covered more than 3300 miles. A Great Circle flight from Chicago to Copenhagen was nearly 500 miles shorter. The Great Circle distance separating Ireland from Newfoundland was nearly 2000 miles, and that was the most direct ocean crossing. But no aircraft existed that could take a payload that far, and in any case the weather over that direct stretch was horrendous at best, with disastrous fogs around Newfoundland and vicious westerly prevailing winds, making a return trip from Europe a continuous battle. If one could leverage the Greenland icecap as a way station, it would be possible, even with mid-1920s aviation technology, to establish shorter, more northerly, and regular passenger-carrying air routes between North America and Europe.

Bert R.J. "Fish" Hassell was a reserve Army Air Corps officer who had earned his wings in 1914. In 1923–24, he had flown for a year in the Canadian bush and based on that experience was hired to deliver an aircraft to a government outpost in northernmost Labrador in 1925. There, a Canadian officer showed Hassell "one of the first semi-polar-projection charts I had seen. On it, the Canadian north appeared as it really is, with major features in their true relative locations and dimensions: The Hudson Strait; Hudson Bay; the passage to the sea at the southeast end of Frobisher Bay; and the port at Churchill, Manitoba, which was the rail end of Canada's great grain-growing areas" (Hassell, 1987: 58).

Hassell immediately foresaw the possibilities in the development of a short hop, payload-carrying air route across Greenland. "A few more developments, a little more experience, and Greenland ice mountains will be no more of an obstacle than were the mountains of Pennsylvania or the Rockies of the West" (*ibid.*: 59). Hassell was empowered in his dream by a fortuitous meeting with University of Michigan geologist William Herbert Hobbs. Hobbs would eventually lead four major expeditions to Greenland and by the mid-1920s was the American expert on the island's geology and glaciology. In the summer of 1928, Hobbs would be leading his third expedition to Greenland, and he proposed to employ his students to clear a runway for Hassell's aircraft so that the aviator could attempt a direct Great Circle flight from Rockford, Illinois, to Hobbs' camp tucked inside of Søndre Strømfjord (Kangerlussuaq).

All Hassell needed now was an aircraft specially fitted for extra fuel, and a willing co-pilot to share the piloting and navigation. He found his aircraft in Detroit, Michigan, in the workshop of Edward Stinson's recently founded Stinson Aircraft Company. The Stinson SM-1 Detroiter was a high wing monoplane soon to be in regular airline service. Hassell's long-range model was equipped with a Wright Whirlwind J-5 engine and modified with the passenger seats removed to allow the addition of a 700-gallon fuel tank. A crash landing during a test flight forced Hassell to add refueling stops to the planned Great Circle route flight to ease the take-off load on the aircraft.

Hassell then set about securing Danish permission to land in Greenland, as well as Canadian permission to fly over northern Canada, and U.S. government permission to undertake the flight at all. The Department of Commerce was not keen, as by 1928 there had been numerous such attempted flights which they considered stunts and many of these had ended in disaster. Hassell got around the bureaucracy by essentially hiring one of them, a 5′5″ ball of enthusiasm by the name of Parker "Shorty" Cramer, a Commerce Department aeronautical inspector. Cramer and his brother William had scratched out a living in the 1920s as barnstormers, flying under bridges, making barely survivable parachute jumps, and selling war-surplus parts for Curtiss JN-4 Jennys. Inspired by the commercial possibilities of a transatlantic air route that covered more land than ocean, Cramer had begun devoting all his time to its development and exploitation.

The difficulties encountered establishing a North Atlantic air route were enormous. No infrastructure existed to handle land aircraft or support them once on the ground, and where open water existed for flying boat operations there were no marked channels, no channel sweeping equipment, no loading gear, and no accommodations for pilots or passengers. If the Hassell-Cramer project plan had one obvious flaw, that was it. It took little account of the fact that the aircraft itself was only one aspect of any air operation.

Both men understood this, but, without a financier behind him, were powerless to change it. The kind of blue water flying boat that could span the Atlantic would not roll out of Igor Sikorsky's workshop until 3 years after Shorty Cramer had vanished over the North Sea. They were attempting a trial concept that would establish a regular transatlantic airline using the only technology and geography that existed for them to do so in 1928.

Cramer sought construction engineers who could build an infrastructure where it was needed most, above 55 degrees north, just north of some of the worst climatic and weather conditions in the world. Only as late as April 1928, had an airplane crossed the Atlantic from east to west, when the Junkers W33 *Bremen* made the crossing from Ireland to Greely Island, off Labrador. As Cramer himself wrote: "The proposed route is north of the dangerous fog and storm area of the North Atlantic [and] by crossing Greenland at two different levels it will be possible to come in with a tail wind at high levels and go out at lower levels with a tail wind" (Cramer, 1929a) (Fig. 7.3).

The compromise plan was to make short hops that were laid out to provide maximum flexibility in case the aircraft had to make a forced landing. Given the take-off restrictions, the pair would now fly first from Rockford, Illinois, to Cochrane, Ontario, a distance of some 800 miles. On August 16, 1928, with 250 gallons of fuel on board, the *City of Rockford* took off not east, but north, and reached Cochrane after a smooth flight of less than 7 h. There, William Cramer had arrived earlier and deployed the local Canadians to chop away stumps and stamp out a 4400-foot runway in the airport-less town. After a night's sleep, Hassel and Cramer lifted off for what was expected to be a 24-hour flight towards Professor Hobbs' camp on the shore of Søndre Strømfjord on the west coast of Greenland.

The fliers enjoyed a flawless flight to Cape Chidley at the northern tip of Labrador and over the small outpost of Port Burwell, which Hassell remembered well from his previous flight there. Cruising at about 110 mph at 5000 feet above the surface, they took their final sights for the hop across to Greenland. Burning 35 gallons of fuel per hour, they soon entered a layer of clouds both above and below them. Hassell considered descending below the cloud layer but was stopped by a warning he had received prior to the expedition. "I recalled what a bright young U.S. Coast Guard officer, Lt. Cdr. Edward H. "Iceberg" Smith, had told me. While I was at Burwell with the RCAF, Smith's cutter was standing offshore, and I talked with him. He said that usually in August, icebergs broken from the Greenland glaciers, as well as Labrador shore ice, tumbled and rolled about until stabilized. Then they poked their jagged peaks into cloud layers, sometimes 300 feet above the sea" (Hassell *op cit.*: 84).

As they crossed the Davis Strait for 4 h and closed on the west coast of Greenland, Hassell and Cramer went over what they had learned from Professor Hobbs. "No ivory tower observer, he had visited Greenland and spent much of his lifetime observing, sampling, appraising, and analyzing Greenland weather... Hobbs felt there was, among other things, a general lack of understanding of Greenland's influence on weather in Europe" (*ibid.*). Air currents off the great ice cap descended at various speeds westwards towards the western fjords, and eastward towards the eastern fjords. The *City of Rockford*, if it could locate Hobbs' camp, would be landing into the wind coming down off the ice.

Prior to the flight, Hassell had sent his mechanic, Elmer Etes, to stay with Hobbs and clear a 2000′ runway along the fjord by Hobbs' base. Hobbs and his students found a likely site about 6 miles up the fjord from the camp, where the fjord shallowed. As he later wrote: "We were ... able to lay out an airstrip lengthwise of the

Fig. 7.3 Fish Hassell and Shorty Cramer alongside the *City of Rockford* in Cochrane, Ontario, after the first leg of their epic flight to the Greenland ice cap (courtesy Midway Village Museum)

fjord and more than a quarter of a mile in length. This was clear of both cracks and boulders..." (Hobbs 1952: 119) (Fig. 7.4).

In addition, Hobbs had had shipped from Copenhagen 400 gallons of aviation fuel. This his students moved from the camp to the airstrip so Hassell and Cramer could refuel the *City of Rockford* prior to crossing the ice cap on their way to Reykjavik, Iceland. The cans were laid out as an arrow pointing towards the landing

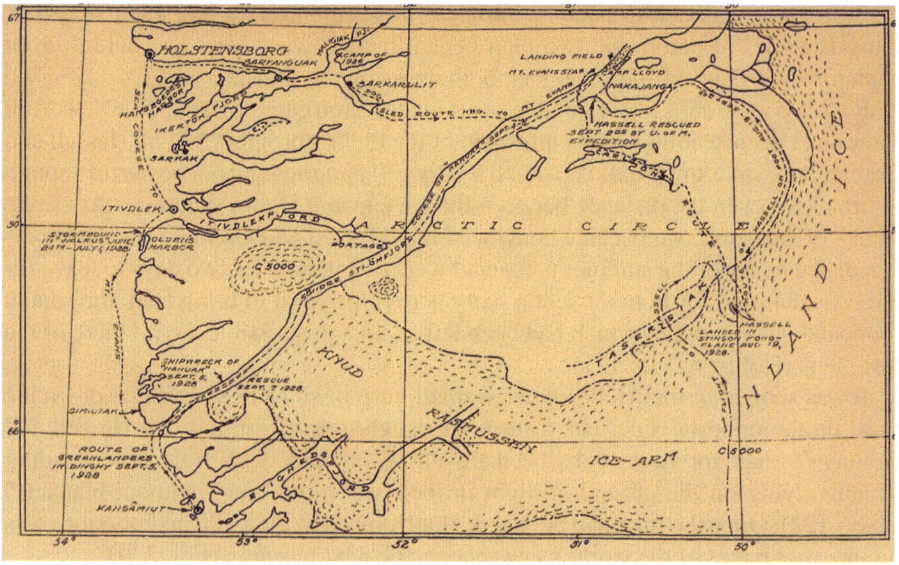

Fig. 7.4 Map of Hobbs' camps in 1926 and 1928 and the location where the Stinson made its forced landing (from Hobbs 1930)

field. However, on the date of their expected arrival, Hobbs and his team listened in vain for any sign of the aircraft over their radio.

Just prior to their arrival at the Greenland coast, the cloud cover cleared and Hassell and Cramer viewed Greenland for the first time, "blinding-white reflected sunlight as it flashed across the great Greenland ice cap… As Dr. Hobbs had predicted, we found the west coast of Greenland clear of bad weather" (Hasell *op cit.*: 85). Unfortunately, the aircraft was almost out of fuel and neither Hassell or Cramer could be sure just which of the many fjords ahead was the one that led up to Hobbs camp. Hassell thought he could see Mount Evans near Hobbs camp about 30 miles ahead, but there was not enough fuel to get there. There was no possibility of landing along the coast without wrecking the aircraft. The fliers made the only choice possible, turning northeast and rising towards the ice cap, where Hassell made a perfect landing as Cramer tapped out the call signal of the aircraft, a signal in the end no one heard.

Hassell let the engine tick over until the last drop of fuel was exhausted and then shut it down. The two men thought they would simply walk down to the coast in a day or so, sledge the fuel up to the Stinson, and be on their way. "Shorty and I so completely misread our predicament that we scooped up handsful of snow, packed then into snowballs, and pelted each other" (*ibid.*: 87). It was then that they heard and felt their first ice tremors. The 'solid' ice cap was in perpetual motion, "a living thing, moving and treacherous" (*ibid.*). They had a rifle, a Swedish axe, and a 10-lb. bag of pemmican bought as an afterthought before the flight. Cramer had a caribou skin parka and, in the end, it saved both men. Cramer wore it during the day and at night the men sat back-to-back and draped it over themselves like a tent. The last thing Hassell did before the two men began to walk west—and he couldn't even explain why to himself—was to lock both doors on the Stinson.

For his part, Hobbs heard nothing overt his research camp's radio. The New York Times had been keeping Hobbs informed of the flight's progress, but all Hassell and Cramer had was a small radio operated by dry cell batteries and not powerful enough to broadcast even the distance between the ice cap and Hobbs's station. "After two weeks of suspense, we became convinced that Hassell and Cramer must have perished, and as we of the summer personnel were due in another two days to leave for the coast enroute for home, I sent a party across the fjord to bring back the canoe, food, and other supplies which had been left against the possible arrival there of the lost men" (Hobbs *op cit.*: 121).

It was soon after that Hobbs noted a small wisp of smoke rise from far down the fjord on the opposite side. His team immediately set off, to find that Hassell and Cramer "...had for two weeks been travelling on foot, with little food, wading streams, lying on the ground at night in their wet clothes and without blankets" (ibid.: 122). Once the men had been fed, Hobbs raced to the radio and his story was on the front pages of the world's newspapers the next morning (Fig. 7.5).

Speaking after the expedition to the Explorers Club, Shorty Cramer thought that the epic flight had exposed the fundamental weakness of their plan. "...this route is not adapted to land equipment...

> (Only) seaplanes and amphibians would provide the most efficient medium of transportation. From the time we left the United States until we arrived at Greenland there was scarcely a place where we could have made a successful landing and even though this might have been possible in the few instances, it would have been practically impossible for us to take off again had we made such a landing. The country was literally covered, though, with lakes and rivers, any one of which would have made a suitable landing place in the event that we were forced down in a seaplane or amphibian (Cramer, 1929b).

One person who already understood Shorty's predicament was also one of the few people in the U.S. prepared to help him overcome it. Early in 1929, Cramer approached 'Colonel' Robert McCormick, owner of the Chicago Tribune, who was about to purchase two such amphibian aircraft from Igor Sikorsky. *Arf Pint* and *'Untin' Bowler'* were both Sikorsky S-38 amphibians, costing over $50,000 apiece in 1929, and the Colonel preferred them for their ability to alight on lakes in the wilds of Canada, where McCormick owned both vacation hideaways and the pulp mills that produced the paper for his publishing empire.

With the popularity of aviators and their exploits at an all-time high after the 1927 transatlantic epic of Lindbergh, it seemed only natural to McCormick to combine his own aeronautical passion with that of the public. In Parker Cramer, he had a famous aviator who had held the world stage for 2 weeks in August and September of 1928 during his disappearance on the Greenland ice cap. McCormick adopted Cramer's plan as his own: to send one of his Sikorsky S-38s off on the uncharted Great Circle Route to Europe, making it the first aircraft to fly round-trip from the New World to the Old and back again.

NEW YORK, MONDAY, SEPTEMBER 3, 1928. TWO CENTS

HASSELL AND CRAMER RESCUED IN GREENLAND; FLIERS WANDERED TWO WEEKS OVER ICY WASTES; SMOKE SIGNALS SEEN, SAVED BY HOBBS EXPEDITION

WHERE THE RESCUED FLIERS CAME DOWN.

Map Shows the Route Traveled by the Plane Greater Rockford Till Its Descent Two Weeks Ago Yesterday on the Sukkertoppen Ice Arm. The Distance Between That Point and the Coast on the Fjord Opposite Mt. Evans Where They Were Rescued Was Covered in Their Two Weeks of Wandering. The Unshaded Line Shows the Projected Route of the Fliers From Which They Were Driven by Storm.

LIVED ON 8 OUNCES OF PEMMICAN A DAY

Eskimos Gave Clue--Rescuers Sent Two Boats After Castaways--Flashlight Signaled "Hassell Safe" to Camp.

LANDED THE PLANE UNDAMAGED ON AUGUST 19

Driven Off Course by Storm and Running Out of Gasoline, They Were Forced Down on Sukkertoppen Ice Arm After Being Sighted.

By PROFESSOR W. H. HOBBS,
Leader of the University of Michigan Greenland Expedition.
Copyright, 1928, by The New York Times Company.
Wireless to THE NEW YORK TIMES.

MOUNT EVANS, Greenland, Sept. 2.—Bert Hassell, pilot of the plane Greater Rockford, and his co-pilot, Parker Cramer, missing for two weeks after they had started on the second leg of their flight from Rockford, Ill., for Stockholm, Sweden, are in Camp Lloyd, safe and well.

They arrived in our motor boat at 9:15 P. M. (Greenland Time) after we had found them on the south side of the fjord in response to a smoke signal they sent up.

They landed on the Sukkertoppen ice arm from the inland ice on Sunday [Aug. 19] after being seen by the people of Fiskernaes.

Traveling for Two Weeks.

Hassell states that the airplane is undamaged. For two weeks the two men have been traveling, first on the crevassed ice surface and later in the wild, unexplored country south of here. They have had a pemmican ration of eight ounces once a day and, though hungry, are in excellent condition.

When I left them at Camp Lloyd an hour ago to send this message they were doing justice to our thick soup and caribou steaks.

Eskimos Saw Smoke of Fliers.

Their discovery today is due to the arrival at our camp of an umiak of Eskimos who, taking their course near the other side of the fjord, six miles distant from here, saw a small smoke near the shore and reported it to us.

As all the caribou hunters have already gone out of that district it was clear that the fliers must be there.

ITALIA HOPE REVIVED BY SMOKE IN ARCTIC

Fishing Smack Reports Campfire on Edge Island, Spitsbergen, a Fortnight Ago.

ICE PREVENTED APPROACH

Norwegians Connect Blaze With "Balloon Group" of Dirigible

THE TWO MEN SAVED WHEN HOPE SEEMED GONE.

Fig. 7.5 The world woke up to the rescue of Hassell and Cramer on September 3, 1928

By convincing McCormick to sponsor a Chicago entry into the international air derby, Cramer could accomplish the creation of Chicago as an international city. McCormick and the *Tribune* could pioneer an airmail and passenger route between Chicago and Berlin over the Greenland ice, north over northern Quebec, over Remi Lake, Rupert House, and Port Burwell. On to Greenland and from there to Iceland, Norway, and Germany.

Powered by two 450 horsepower Pratt & Whitney Wasp engines, McCormick's S-38 received special modifications for the planned transatlantic run. The cabin accommodations were removed and replaced by three 100-gallon fuel tanks. On July 3, 1929, the S-38 lifted off from Chicago and, after several improvised stops along the way on account of fog, the aircraft descended on Port Burwell, northernmost Labrador, 6 days later. It wasn't exactly the rapid crossing to Europe Cramer had imagined, and before the crew of three could even lift off for Greenland, ice flowed into the inlet where the aircraft lay floating, punctured the thin aluminum hull, and then carried it offshore where it disappeared beneath Ungava Bay, never to be seen again (Fig. 7.6).

Little more than a month after his return from Port Burwell, Cramer shipped out to Antarctica as the chief pilot for Sir Hubert Wilkins on the Second Wilkins-Hearst Antarctic Expedition. He returned to embark on a lecture tour arranged by his brother, but the two of them never stopped trying everything they could to obtain another seaplane for another attempt on the Northern Route. But times and conditions had changed. While he was on his way to the Southern Continent the stock market had crashed and Cramer returned 6 months later only to find companies unwilling to risk their money on a gamble that had already produced two failures.

The Cramer brothers composed a blizzard of letters to prospective sponsors, and each letter returned with the same bland pronouncements about current economic conditions. Shorty Cramer felt other pressures as well. He had invested a decade in the quest for a Northern Air Route, and now several well-heeled rivals were poised to launch northern expeditions of their own. Gino Watkins' British effort would soon land at Angmagsallik on Greenland's eastern shore, complete with a ski-equipped plane to attempt the still-unachieved first hop over the ice cap. Harvard's Alexander Forbes set sail aboard the schooner *Ramah* in 1931, with two seaplanes aboard, to chart the coast of northern Labrador, the very kind of geographic legwork Cramer knew was essential to the northern route's success. And Cramer's midwestern rival, Charles Lindbergh himself, was planning his own seaplane expedition over the northern route, sponsored by one of America's biggest airlines. As a friend wrote to Cramer: "It seems almost needless to call your attention to the danger confronting your air route should Lindbergh anticipate your efforts... If he should make

Fig. 7.6 The Sikorsky S-38 trapped by ice at Port Burwell, northernmost Labrador, July 1929 (from Capelotti, 1995)

his proposed northeastern flight he would rob you of certainly much of your earned credit..." (Jones to Cramer, 1931).

By June of 1931, Cramer accepted an offer from Transamerica Airlines, a small midwestern feeder line, to pilot their *Survey Plane No. 1*, a small Bellanca seaplane, across the Northern Route. He lifted off in early August 1931, with a Canadian radio operator, Oliver Pacquette, and accomplished what no pilot had done before: the first west-to-east crossing of the Greenland ice cap. He then continued on from Angmagsallik to Reykjavik, then to the Faroes and then the Shetland Islands. From there, he had but one final short 200-mile hop to the coast of Norway, then on to Copenhagen and glory.

The seaplane left Lerwick in the Shetland Islands on Sunday morning, August 9. At 12:30, Paquette radioed that they had sighted the Norwegian coast and were fighting a heavy storm, but they expected to arrive in Copenhagen at 4:30. It was not until 5 weeks later, on September 17, 1931, when the Bellanca was finally recovered, floating near Scotland. Neither Cramer nor Paquette was on board and neither was ever found.

Just 6 months later, after the Atlantic had been flown by a man (Lindbergh), and a woman (Amelia Earhart), and claimed the lives of Cramer, Pacquette, and a host of others, it was one George R. Hutchinson's idea to fly the Atlantic with his entire family. Undeterred by the adverse press reaction to the announcement of his plans, Hutchinson, a pilot himself, bought an S-38, named it *City of Richmond*, and recruited a navigator, radioman, mechanic and a cameraman.

The entourage departed from Floyd Bennet Field on Long Island on August 23, 1932, and flew to St. John, New Brunswick, the same day. There, to save weight and increase the fuel for the overwater legs of the journey, the 500-pound set of wheels were removed to make the S-38 a true flying boat. The next day the expedition flew to Anticosti Island in the Gulf of St. Lawrence, where they waited out 6 days of bad weather. After the skies cleared, the S-38 reached Hopedale, midway north on the Labrador coast. So far Hutchinson had done as well as the Chicago *Tribune* flight of 1929.

On September 2, against the direct orders of the Danish Government (which forbade Hutchinson from landing anywhere on the coast of Greenland—effectively stopping the transatlantic flight), the S-38 flew the 600 miles of the Labrador Straits and alighted in the harbor at Godthaab, Greenland. The Danes responded by slapping a punitive fine on Hutchinson, but not one that in any way stopped him. He continued on his way around the tangled coast, forsaking an attempt to cross the ice cap, and reached Julianehåb harbor, near Greenland's southern tip, on September 7.

So far, Hutchinson had every reason to be proud of himself. Several more experienced pilots on the Northern Air Route had turned back or been forced down before reaching the point where he now stood. But Hutchinson was about to learn what they already knew—that Greenland had to be crossed via the ice shield, much farther north, and not around the perpetually fogbound southern coastline.

As they left Julianehåb *en route* to the small settlement of Angmagssalik on the eastern shore, fog closed in on the S-38. The treacherous shoreline soon disappeared into the mists. One of the S-38 s fuel tanks sprang a leak. Hutchinson could go neither forward nor back. He brought the aircraft down at sea, 50 miles from Angmagssalik, and began taxiing for shore. The radioman sent out an SOS, and an immediate search was begun.

The S-38 began to take on water as it neared shore, so Hutchinson rammed the flying boat onto a rock outcropping, and all hands abandoned ship. They endured that night huddled together on the rocks, and in the morning light watched the swamped S-38 wallow amid the drifting ice. A fishing trawler spotted the marooned family the next day and the Hutchinsons and their flight crew were rescued (Fig. 7.7).

The definitive aerial work on the Greenland route was yet to come. In 1933, with Shorty Cramer lost at sea, Charles Lindbergh cleared the field of amateurs like the Hutchinsons and with his wife Anne Morrow Lindbergh set out on an extended series of survey flights across the North Atlantic via Greenland. Their Lockheed Sirius monoplane was built to Lindbergh's specifications and was equipped, as Shorty Cramer's experience had shown, with floats instead of wheels, and a more powerful 710 hp. Wright Cyclone engine. All empty spaces, including the floats were used for fuel storage. The Lindberghs were sponsored by Pan American Airways, which provided a fully equipped support vessel to follow the couple and serve as a floating repair base for the Lockheed (for the full details of the flight, see Anne Morrow Lindberg, 1934, and Anne Morrow Lindberg, 1938).

Fig. 7.7 The Flying Hutchinson's watch as their Sikorsky is dashed upon the shore near Angmagsallik (Tasiilaq), East Greenland (from Capelotti, 1995)

The couple departed the U.S. in early July and arrived in Cartwright, Labrador, on July 14, where the Pan Am expedition ship *Jelling* was waiting for them. The trans-Atlantic air route remained front page news, with the arrival of the Lindberghs in Cartwight and the departure just the day before of the Italian Aerial Atlantic Cruise, an armada of two dozen Savoia-Marchetti S. 55X flying boats led by Italo Balbo.

Fog and rain delayed the Lindberghs in Cartwright for a week, while the *Jelling* proceeded ahead of them to the west coast of Greenland. Fog eventually rerouted their 'direct' flight to Greenland north along the Labrador coast to the mission settlements of Hopedale and Hebron. They finally took off from Hebron on the afternoon of July 22 and little more than 6 h later landed at the harbor of Godthåb (Nuuk), the capital of Greenland. "We were three hours out from the coast of Labrador when we saw them... First we could only see their tops, like clouds on the horizon, then a jagged blue line, and finally, as we came nearer, white caps and white-streaked sides stretching as far as the eye could see... Not until we were quite near did we see something that looked like green moss in the steep valleys" (Lindberg, 1934 *op cit.*) (Fig. 7.8).

Fig. 7.8 Detail from Charles Lindbergh's annotated map of the 1933 survey flights around and across Greenland (National Air and Space Museum, Smithsonian Institution)

As they were towed to the government wharf, kayaks surrounded the Lindbergh's aircraft. "Coming nearer, we could see the Greenland women all dressed in holiday style—high red boots, sealskin pants, bright blouses, and tall knitted caps.

> As we stepped from the boat to the dock they all shouted and followed us up the street to the Governor's house, where the Governor and other members of the colony gave us a hearty Danish welcome. Here we met Commander Dam, the representative of the Danish Government who was to accompany us on the *Jelling*, and who gave us much assistance during our flights in Greenland... The Greenlanders (they are always called Greenlanders here and not Eskimos) that we met as we walked around were prosperous looking. Healthy and gay, they were in their best clothes, for it was Sunday and therefore no one was working. The long kayaks were pulled up on shore and fish were drying on overturned boats. We saw a few of the original style of Eskimo huts, made of turf and some bricks. But even these had windowpanes, through which, in one case, I could see a flower growing in a can, a needle and thread, and a cup of coffee (*ibid.*).

The Greenland aerial survey flights proper began on July 25, with a flight to Holsteinborg (Sisimiut), where the Lockheed circled above Professor Hobbs' camp, and climaxed with a six-hour flight across the ice cap on August 4. At Ella Island on Greenland's east coast, the couple met with Denmark's preeminent geological explorer of Greenland, Lauge Koch, himself an expert in the use of aerial surveys for cartography. Flying from Koch's base camp at Ella Island to his northern camp at Clavering Island, the Lindberghs proceeded down the coast to Angmagsallik (Tasiilaq).

On August 7 they conferred with famed Danish-Greenlandic anthropologist Knud Rasmussen, who would pass in Denmark later that same year. "It was at Angmagsallik we were fortunate enough to meet the late Dr. Knud Rasmussen, the famous explorer, and his mapping and surveying party. Dr. Rasmussen has interpreted Greenland to many people through his books, and wherever he was and among the people who worked with him, there was a remarkable enthusiasm and spirit" (Lindberg, 1934 *op cit.*).

It was at Angmagsallik that Anne Lindbergh asked a young Greenlandic boy to paint on the aircraft the name that the Greenlanders had given it: *Tingmissartoq*, 'one who flies like a big bird.' Anne Lindbergh admitted that it was the enthusiasm of Rasmussen and his expedition that led them to see Angmagsallik as the most beautiful place in all of Greenland. "...it was Angmagsallik that remains in my mind—prickly, flame-like mountains across the fjord, red houses climbing up a green gulley, patches of bluebells halfway up the hill, bright boots hung up to dry in the sun, and Greenlanders, in the evening light, twirling round and round on a dusty path to the whine of an accordion" (*ibid.*) (Fig. 7.9).

After a brief return flight first to Godthåb and then Julianehåb (Qaqortoq), the Lindberghs on August 15 left Angmagsallik for Iceland and the rest of their Atlantic round-trip. After their transatlantic expedition for Pan American, the Lindbergh's donated the *Tingmissartoq* to the American Museum of Natural History, ironically enough given that museum's tortured relationship with Greenland and Greenlanders. The museum gave it to a reluctant U.S. Air Force Museum in Ohio, in the 1950s,

Fig. 7.9 The Lindbergh's Lockheed Sirius *Tingmissartoq* in the Pioneers of Flight Gallery at the National Air and Space Museum of the Smithsonian Institution, September 11, 2025 (photo by the author)

and it eventually wound up in the Air and Space Museum of the Smithsonian Institution in Washington, D.C., as a main feature of the newly redesigned Pioneers of Flight Gallery.

Shorty Cramer's Sikorsky S-38 lay somewhere at the bottom of the coastal seas of northernmost Labrador, while the Flying Hutchinson's S-38 was forever lost off East Greenland. Forty years after the 1928 epic of Cramer and Hassell on the Greenland ice cap, a survey flight spotted their Stinson Detroiter, still awaiting their return. With Shorty long dead and Fish an old man, it was Fish Hassell's son Vic and Shorty's brother William who, along with aviation artist Robert Carlin, organized a daring expedition in 1968 to rescue the aircraft from the ice and return in to Rockford, Illinois, where it was eventually restored and now serves as a main exhibit of the Midway Village Museum. With the return of the *City of Rockford* from the ice, the age of the pioneer aviators over Greenland was well and truly over (Figs. 7.10 and 7.11).

It was now time for American scientists and soon, citizen soldiers. As the first of those scientists, William Hobbs had speculated, as his students cleared a makeshift runway for Cramer and Hassell in 1928: "If we had been supplied with a gang of laborers, or, better, with a bulldozer, we could have prepared a runway for the

Fig. 7.10 Recovering the *City of Rockford* from the Greenland ice cap, 1968 (courtesy Midway Village Museum)

largest bombers" (Hobbs 1952: 119). In fact, his small University of Michigan team had cleared the first runway of exactly that. The *City of Rockford* landing strip would become site of the largest Greenland airfield during the Second World War, Bluie West 8. After the war, it became the Söndreström Air Base and, in 1992, Kangerlussuaq Airport, the largest airstrip in Greenland.

Fig. 7.11 A Greenlandic Airways helicopter lifts the Stinson from the ice cap and delivers it to Söndre Strömfjord air base in 1968 (courtesy Midway Village Museum)

References

ABC News. (2025). *Former Danish minister for Greenland discusses Trump's push to acquire island*. (Paraphrase of Wilson by former Danish minister to Greenland, Tom Høyem). Retrieved January 6, 2026, from https://abcnews.go.com/International/former-danish-minister-greenland-discusses-trumps-push-acquire/story?id=118140297

Bartlett, R. (1940). Greenland from 1898 to now. *National Geographic Magazine, LXXVIII*(1) July 1940.

Capelotti, P. J. (1995). *Explorer's Air Yacht: The Sikorsky S-38 flying boat*. Pictorial Histories.

Convention between the U.S. and Denmark (1917). *39 Stat. 170*. Retrieved January 6, 2026, from https://www.doi.gov/sites/doi.gov/files/uploads/vitreaty.pdf

Cramer, P.. (1929a). *A Northern Air Route* (4), William H. Cramer Collection, courtesy of the Cramer family.

Cramer, P.. Ca. 1929b. *Winging Past the Midnight Sun*. William H. Cramer Collection, courtesy of the Cramer family.

Ernest Jones to Parker Cramer, June 12, (1931). Cramer Collection. Another friend wrote to Cramer that: "People don't seem to take to this sort of thing like they did a year ago, and it is no doubt due to the number of flights made in the past year which have been given so much cheap publicity, and also, because of the financial depression." Jack Ford to Parker Cramer, February 2, 1931, Cramer Collection. There is also a peculiar post-script from a letter from Prof. Hobbs at the University of Michigan, written to Cramer after the latter returned from Antarctica, which reads: "If you are planning to see Col. McCormick better see me first. I have much to talk over with you." William H. Hobbs to Parker Cramer, April 1, 1930, Cramer Collection.

Emmerson, C. (2011). *The future history of the Arctic*. Vintage Books.

Grover, D. H. (2001). Richard E. Byrd and the 1925 MacMillan Arctic Expedition. *Aviation History*: March.

Hassell, B. R. J. (1987). *Fish Hassell—A Viking with wings*. Maverick Publications.

Hobbs, W. H. (1930). *Exploring about the North Pole of the Winds*. New York: Putnam.

Hobbs, W. H. (1936). Peary. New York: MacMillan.

Hobbs, W. H. (1952). *An Explorer-Scientist's Pilgrimage*. Ann Arbor, Michigan: J.W. Edwards.

Lindberg, A. M. (1934). Flying around the North Atlantic. *National Geographic Magazine, LXVI*(3), 259–337.

Lindberg, A. M. (1938). *Listen! The Wind*. Harcourt.

New York Times. (1918). *Peary to direct polar air flight*. December 23, 1918, p. 1,5.

U.S. Department of State Archives. (1901). *The Secretary of State to Minister Swensen*. File No. 111, Retrieved January 6, 2026, from https://history.state.gov/historicaldocuments/frus1917/d672

U.S. Department of State Archives. (1910). *Minister Egan to the Assistant Secretary of State. File No. 711.5914/14*, online at: Retrieved January 6, 2026, from https://history.state.gov/historicaldocuments/frus1917/d759

Welky, D. (2017). *A wretched and precarious situation*. W.W. Norton.

Chapter 8
Captain Mac, Professor Hobbs, Louise Arner Boyd, and Greenland as the Center for U.S Polar Science

Abstract As exploration gave way to field science in the 1930s, expeditions led by William Hobbs, Donald MacMillan, and Louise Arner Boyd produced a massive body of fieldwork that not only laid the foundation of American scientific research in Greenland but would become critical to the military planning for the American defense of Greenland during the Second World War.

Keywords Greenland · Greenlandic Inuit · Bowdoin · Polaris Beach · Etah · Søndre Strømfjord · National Geographic Society · University of Michigan · Meteorology · Cartography · U.S. Army Air Corps · American Geographical Society

> *How pitifully small our own travels and how little we have seen! My imagination runs riot when I think of the spring and fall migrations of the birds. However much we love the North, our life here would be a bit dreary without the birds. We look for them in spring, we enjoy them during the summer, we miss them when the lakes begin to skim and twilight creeps over the land.*
>
> —Donald MacMillan, *Etah and Beyond*, 1927

Robert Peary's disciples continued to stage expeditions expeditions to Greenland throughout the interwar period. Donald MacMillan had a purpose-built ship, the *Bowdoin*, constructed for just such voyages. His June 23, 1923, departure for Greenland from Wiscasset, Maine, was attended and feted by no less than America's greatest living Arctic leader, Adolphus W. Greely, by then a Major General of the U.S. Army on the Retired List. Greely's longevity, along with a lack of accountability for the failures of the Lady Franklin Bay expedition, had placed upon him the mantle of *éminence grise*, first president of the Explorers Club, author, visiting dignitary at the coronation of George V, and in the 1930s only the second recipient of the Congressional Medal of Honor for a lifetime of government services rather than for physical courage at the risk of one's life.

MacMillan, for his part, would return the favor for Greely's less than inspiring "recital of … the hardships of all and the death of most of his party and the crushing

P. J. Capelotti, *Lost Greenland*, Springer Polar Sciences,
https://doi.org/10.1007/978-3-032-27841-8_8

of (his ship)" (MacMillan, 1927: 1–2). After the *Bowdoin's* overwintering in northwest Greenland, MacMillan's crew in May of 1924 would drag a heavy memorial tablet to Cape Sabine on Ellesmere Island. Finding a suitable boulder nearby, the tablet was bolted on, to read: *To the memory of the dead, who, under Lieutenant A.W. Greely, here gave their lives to ensure the final and complete success of the first scientific cooperation with other nations, 1881–1884. Erected by the National Geographic Society, 1923*." Combined with the construction of the massive Peary monument a decade later close across the ice in Greenland, it marked another milestone in the society's self-appointed role as arbiter of who was memorialized in the Arctic and who wasn't (Fig. 8.1).

The expedition established a pattern MacMillan would follow over what became a nearly half century of expeditions and two dozen voyages by the *Bowdoin* in northern waters. A small group of young, unmarried men were recruited with both a particular shipboard skill as well as a level, generous spirit. They then engaged in the kinds of ethnological and systematics collections common to the era.

> The eminent botanist, Dr. Joseph D. Hooker, states that there are about seven hundred and seventy species of plants within the Arctic Circle. In 1953–55, Dr. Kane collected one hundred and six species of flowering plants in North Greenland. In 1860–61, Dr. Hayes added five or six unrecorded ones to this list. In 1913–17, my botanist, W. Elmer Ekblaw, collected one hundred and twenty-five species of vascular plants between Melville Bay and the Humboldt Glacier. It is of interest to note that many of the plants found in Greenland are found in our country only upon the summits of the White and Adirondack Mountains (MacMillan, 1927: 194).

Born in Provincetown, Massachusetts, in 1874, Macmillan passed there 95 years later, by which time the local wharf had been named for him and he had earned multiple nicknames, including 'Captain Mac,' and, in Inuktitut, "Nagelak," the "Leader." He also, at the age of 61, acquired a 29-year-old wife, Miriam, who would accompany him on his later expeditions and together they collected an Inuktitut dictionary to go with thousands of images and audio and motion picture recordings of the Greenlandic Inuit. Miriam would live on until 1987, curating this immense database into what became the Peary-MacMillan Arctic Museum at Bowdoin College in Brunswick, Maine.

It was a far distance from MacMillan's start as an orphan at twelve and, a decade later, a high school teacher with a geology degree from Bowdoin College. A summer school to teach boys navigation and seamanship led to an incident when MacMillan rescued several boat wreck survivors over two nights, a have-a-go hero episode that brought him to the attention of Robert Peary. Peary invited Macmillan on his failed 1906 attempt on the North Pole, and again on his claimed polar trek in 1909. Thereafter MacMillan centered the rest of his life on the people and places of northwest Greenland.

The 1923–24 expedition very nearly followed in the wake of the lost American ships of Kane and Hall and Greely when the *Bowdoin* attempted to break out of its winter ice cradle in Refuge Harbor near Littleton Island in July of 1924 and was almost immediately hung up on a rock. In a harbor with nine-foot tides, the

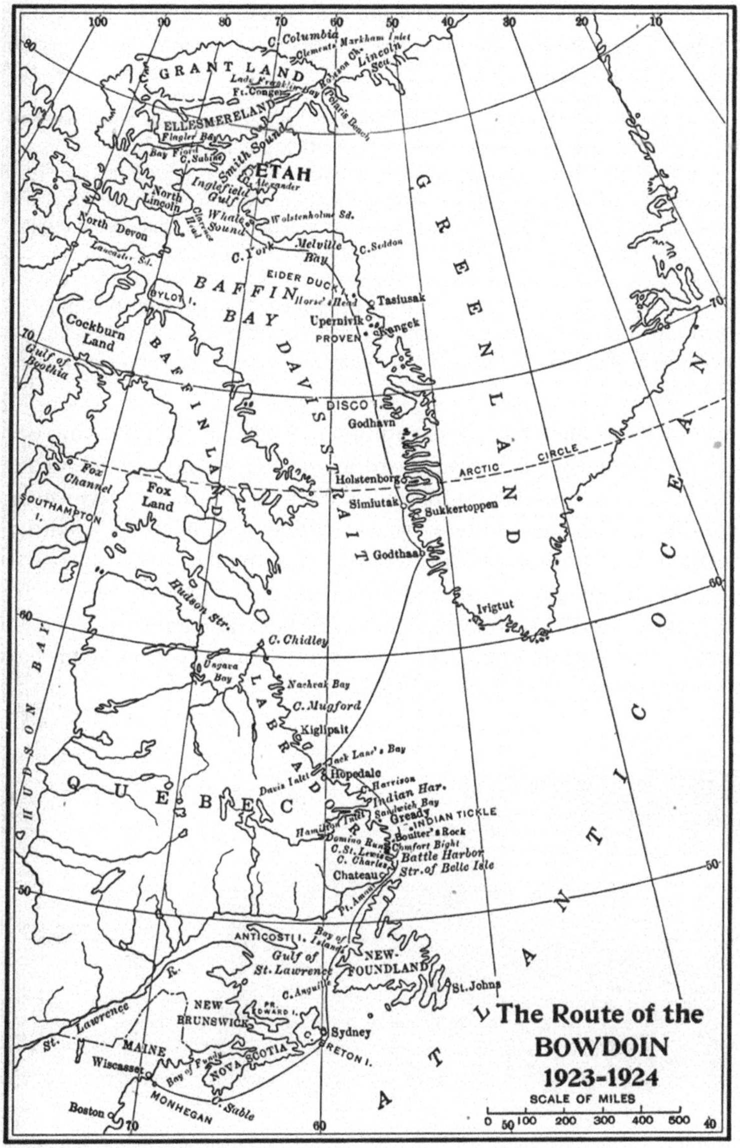

Fig. 8.1 Route of Donald MacMillan's *Bowdoin* during its overwintering in Greenland in 1923–1924 (from MacMillan, 1927)

grounding could have been a catastrophe, with seven men trapped in a remote spot without provisions to endure a second winter. The crew had just until the next flood tide to remove everything they could from the ship, seal all portholes and caulk the doors, and then watch as the tide crept back in and slowly lift the *Bowdoin* from the rock upon which it had stuck (Fig. 8.2).

MacMillan also engaged in arguably the first quasi-systematic historical archaeology of polar exploration, when he sailed a dory to the site of the grounding of Charles Francis Hall's *Polaris*.

> Upon our right hand was Polaris Beach, where the U.S.S. *Polaris* was beached in 1872 in a sinking condition following a bad squeeze off Littleton Island. The ship was abandoned and later drifted from her position and sank between Littleton Island and the mainland. The men constructed a dwelling-house from parts of the ship, lived here during the winter, built two boats in the spring and retreated south. We landed to see what might remain after fifty-one years. The site of the house was littered with bits of wood, coal, glass, nails, screws, door locks, ringbolts, and mast fittings. One brass door button alone was marked 'U.S. Navy 1870.' I had previously picked up many relics here marked 'Washington Navy Yard, 1870.' We marveled at the state of preservation of what was left after so many years, and especially at the lack of rust, a testimonial to the dryness of the Arctic air (*ibid*: 207–208).

Perhaps MacMillan's most famous expedition to Etah was the following year, when he somewhat reluctantly carried on board a small U.S. Navy aeronautical survey detachment led by the infinitely ambitious Richard E. Byrd. The aeronautical work using small float planes was a failure, but MacMillan continued his scientific studies of both the natural and cultural histories of the area. In all his work, he was aided by Danish government officials. "With very few exceptions I have always found the Danes unfailingly courteous and very likable. If the whole nation is to be judged by the character of the Danish officials in Greenland, then we as Americans have much to learn about hospitality to strangers" (*ibid*: 224).

Fig. 8.2 Bowdoin on the rocks, July 1924 (from MacMillan, 1927)

Another Peary expedition veteran who made a lifetime of visits to Greenland was the Newfoundland captain Robert Bartlett. It was Bartlett who transported the materials required for the National Geographic Society to construct its largest monument in the Arctic, the giant Peary obelisk at Cape York. Barlett's specialism was the transport of young boys from preparatory schools in the U.S. on board his schooner, the *Effie M. Morrissey*, to the island for an Arctic summer of collecting "data and specimens for the Smithsonian Institution" (Bartlett, 1940). He also used these occasions to both castigate Greenlanders for abandoning their "old ways" of skilled Arctic hunters in exchange for what he saw as their modern, tuberculosis-ridden state, living inside "houses of wood... (with) the blare of the radio and the monotonous droning of the gramophone.... Then came firearms. One result was the destruction of herds of animals—needless and wasteful destruction. Initiative, independence, the old upstanding qualities of the Peary Eskimo, seemed to pass (Bartlett, 1940). That Peary himself made a career of trading Western goods in exchange for Greenlandic labor (including the removal to New York of their sole source of local iron) entirely escaped Bartlett's paean to Peary and 'his' Eskimos.

Besides using teenaged labor to obtain Arctic collections, Barlett also financed his Greenland voyages through the regular shooting and trapping of Greenlandic wildlife to fill orders from zoos in America. That he and his crew took especial delight in this is evidenced by a photo reproduced in *National Geographic Magazine* showing his crew, posing gleefully behind a polar bear cub they had forcefully separated from its mother and then strung up on the deck of the *Morrissey*, to writhe in agony as the alien vessel readied to transport it "for the Philadelphia Zoo" (*ibid.*). Similar images of polar bears and their cubs shot, killed with bows and arrows, or captured to be cruelly strung up and mocked were published in the children's book *David Goes to Greenland* (1926).

Perhaps the only notable benefit of Bartlett's Greenland voyages was the ship of opportunity he provided to the University of Michigan geologist William Herbert Hobbs as Hobbs sought to establish a research station on the west coast in the summer of 1926. The *Morrissey* had been chartered that summer for another museum collecting and shooting expedition, this one led for the American Museum of Natural History by wealthy publisher George Palmer Putnam.

Putnam had taken his thirteen-year-old son David on board the *Morrissey* a year earlier on a visit to East Greenland, and it was he who published *David Goes to Greenland* later that same year. "The American Museum of Natural History in New York wanted some things from the North for its new Hall of Ocean Life, as well as Arctic birds. So Dad said he would organize an expedition and get the specimens they wanted. Among these are Narwhal, Greenland Brown Shark, walrus, all kinds of seal and many birds... I've a Newton 2.56 rifle and a twenty-two rifle and I hope to get a chance to do some shooting..." (Putnam, 1926: 4–6). David's father, however, is best remembered as the husband and publicist of Amelia Earhart at the very moment Earhart was achieving global fame through her transoceanic flights. The *Morrissey* that summer brought both Putnam and Hobbs to Greenland, as well as Knud Rasmussen carrying resupplies for his research base at Thule.

William Hobbs had been something of a child prodigy in drawing but was ultimately to find renown as one of the premier geological explorers in American scientific history. His professors at both Johns Hopkins and Harvard recognized his geological acumen immediately and offers for year-long study tours of European geological formations became almost a routine for Hobbs, as was a summer position as a surveyor for the U.S. Geological Survey. As a young assistant professor at the University of Wisconsin, he attended the landmark seventh International Congress of Geology in St. Petersburg, Russia, in 1898, a conclave of the most well-known researchers in the field from around the world. The Swedish delegation alone sent a roster that included Otto Torrell, A.G. Nathorst, Arvid Högbom, and Gerard De Geer.

But Hobbs did not enjoy either G.P. Putnam's fame or his money. After he moved on to a professorship at the University of Michigan, financing for what became four University of Michigan expeditions to Greenland between 1926 and 1933 was cobbled together from the usual assortment of grants, including the Charles Stewart Mott Foundation's very first international and environmental grant ($500) for the third expedition in 1928 (see: Mott.org, 1990). Hobbs eventually raised $85,000 to support his Greenland research at the meteorological station he established at the head of Søndre Strømfjord (Hobbs, 1952: 104). By the time of the fifth expedition in 1931, Hobbs had become famous for his Greenland work and stepped back into a stateside advisory role, occupied as he then was by writing up the scientific results of his Greenland work, as well as a popular account, while making time to sit for a portrait bust to be added to those of other notable American scientists at the U.S. National Museum (*ibid*.: 134) (Fig. 8.3).

The first Michigan expedition was a simple reconnaissance for Hobbs and his students, "to find a suitable base and otherwise learn how to outfit the later expeditions." They found a base within a rock-walled amphitheater Hobbs named University Bay, about forty miles east of the coastal colony of Holsteinborg (Sisimiut).

> The objects of these (later) expeditions were two. The first of these was to explore the glacial anticyclone in the upper air by means of pilot balloons and *ballons sondes*, which latter carried meteorological instruments to be recovered later. The other main aim was to see how a continental glacier during its liquidation modifies the land surrounding it, knowledge vital to the interpretation of ancient glacial deposits. Our planet offers but two localities where this could be studied to advantage, for the Antarctic as then explored was everywhere bordered by the sea, and the same is largely true of Greenland as well. The exceptionally wide land border areas are Peary Land in remote northeast Greenland, then almost inaccessible, and the southwest coast, where beneath the Arctic Circle the bordering ribbon of land is more than a hundred miles in width. This locality I selected for my studies, and the choice was fortuitous, as events were to prove (*ibid*.: 104–105).

Those events included the rescue of the pilots Hassell and Cramer at the end of the 1928 field season (Fig. 8.4). They also included the introduction to serious scientific fieldwork of several of Hobbs' students who themselves would make

Fig. 8.3 The supremely self-assured Hobbs sketched in his Greenland gear (source: William Herbert Hobbs papers, Bentley Historical Library, University of Michigan, via Creative Commons license)

significant contributions to both the study of Greenland and its defense from Nazism during the Second World War. One of these students was Max Demorest, the young glaciologist from Flint, Michigan who would go on to develop a theory of glacial ice crystallization and re-crystallization that became widely accepted as the basis of glacial flow. Demorest would lose his life on the Greenland ice cap during the war.

Another student was William S. Carlson. Born in Ironwood, Michigan, Carlson earned all of his academic degrees in geology studying at the University of Michigan under Hobbs. He participated in the 1928 expedition, overwintering in Greenland through 1929, and leading the 1930–31 expedition as a fellow of the Scandinavian-American Foundation. Carlson was on hand for the rescue of Hassell and Cramer but admitted if it had been left to himself and his fellow students, the pilots would likely have been lost as "our summer party was packing up to go home to Michigan.

Fig. 8.4 Gould, Hobbs, and Belknap sending up a weather balloon. Gould and Belknap would narrowly escape drowning during the expedition, saved by native Greenlanders in a kayak and *umiak* (source: University of Michigan News and Information Services Photographs, Bentley Historical Library, University of Michigan)

> "This was during the caribou hunting season, and there were Greenlanders in our camp who had crossed the fjord from the Sukkertoppen hinterland, where they had been hunting, to barter with the college students who were eager to get souvenirs to take home. One of the Greenlanders saw smoke across the fjord; he insisted it could not be from a fire lighted by the hunting party.
>
> We sent over a motorboat to investigate. It brought back Bert Hassell and Shorty Cramer, bearded and grimy, gaunt after fourteen days of near starvation… Just one day later their fire might not have been seen, for the Greenlanders with their keen hunters' eyes would have been gone, and our main expedition would have sailed. The three of us who were to stay at the station would probably have been so busy with our scientific duties that we would not have seen the smoke (Carlson, 1962: 14–15).

Carlson became a close friend of Shorty Cramer and advised him before the 1929 Sikorsky S-38 flight that ended in northern Labrador. The Michigan research station's radio in Greenland was down for most of the 1931 field season, when Cramer successfully flew across the North Atlantic via the Ice Cap. It was only later that Carlson learned that Cramer had wanted him to serve as his support chief in eastern Greenland (Fig. 8.5).

When Carlson and Demorest landed in Copenhagen at the end of their field season, "Peter Freuchen lost no time seeking me out.

> "That was my first meeting with the gigantic, bearded, one-legged Viking whose battered features later became so familiar to countless millions of television viewers, and to the legions of readers of his books about the Arctic. Peter was then a younger Viking, already a

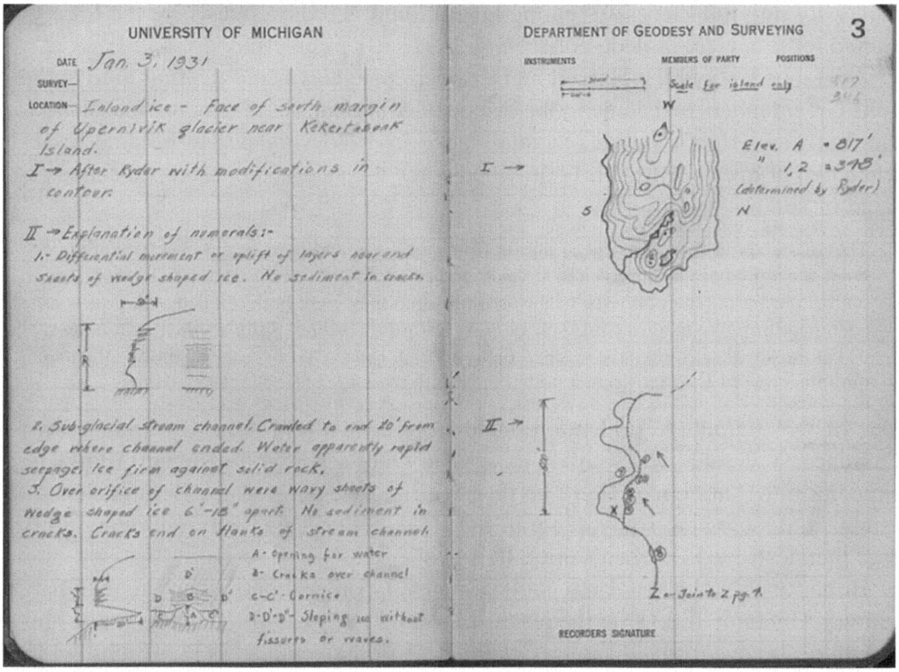

Fig. 8.5 Pages from William Carlson's 1931 Greenland field notebook. Source: Arctic: Surveying Notes, 1931, Box 19. William S. Carlson Papers, 1924–1976, UM 14 (courtesy of the Ward M. Canaday Center for Special Collections, University of Toledo. See also: utoledo.edu (n.d.))

> veteran of the north country, an expert on Eskimo life, and the possessor of tremendous enthusiasm for Shorty Cramer's dream. Any idea I may have had to return to the University of Michigan to take up the classes and the academic routine that awaited me was ruthlessly brushed aside by Freuchen...
>
> Rapid-fire, in that ferocious growl with which Freuchen was accustomed to speak English, I was told the details. When Cramer reached Copenhagen, I was to fly back with him to Angmagssalik.... When I was safely in Greenland, Cramer would return to Denmark, pick up Freuchen, and fly back over the route to the United States (*ibid.*: 18).

When Cramer was lost in the North Sea *en route* to Copenhagen, Carlson was already at the airfield awaiting his arrival. The loss of Cramer and his Canadian radioman Oliver Pacquette meant for Carlson a return to his studies of the Greenland ice. His intimate knowledge of both Greenland and Shorty Cramer's northern air route (the final University of Michigan Greenland expedition would be co-sponsored by Pan American Airways) would make him invaluable to the U.S. Army Air Corps as it hurriedly developed Greenland air bases at the start of the American participation in the Second World War. After the war, returning to academia, Carlson went on to the presidency of three different state universities, before a final fourteen-year tenure as president of the University of Toledo from 1958–72, where he shaped the university into the successful public institution it is today, and where the library is named for him (see: utoledo.edu).

At Hobbs station, the Hassell and Cramer episode was not the first time indigenous Greenlanders had come to the rescue of Americans experiencing the island for the first time. Hobbs initial reconnaissance in 1926 very nearly became his last when two of his students nearly drowned when their outboard-equipped canoe capsized in Søndre Strømfjord.

> They were about three-fourths of a mile from our base, and their plight observed at once, but it was only because an Eskimo kayaker and an *umiak*, or large woman's skin-boat, were both at our base—the only visit of the summer--that they were both rescued. All of us were soon paddling like mad in some sort of craft toward the scene of the accident, but none of us but the kayaker could have reached them in time. He arrived first and supported the men until the crew of Eskimo women pulled them into the *umiak* (Hobbs, 1952: 106).

One of those graduate students rescued from drowning by the Greenlanders was Laurence M. Gould, who went on to serve as chief scientist for Richard Byrd's first expedition to Antarctica. There he flew with Norwegian aviator Bernt Balchen to establish a geological study camp in the Rockefeller Mountains, surviving a blizzard that destroyed Balchen's aircraft.

Hobbs averaged a scholarly publication every other month for more than half a century. His fame following the rescue of the *City of Rockford* aviators made him a sought-after voice on all matters polar. In mid-November 1937, Hobbs was in Cresson, Pennsylvania, invited to give the dedicatory address at the unveiling of the statue of Robert Peary in Peary's birthplace. And while his theory of a fixed single glacial anticyclone atop the Greenland ice cap was discounted in his own lifetime (see, for example, Matthes and Belmont 1950), Hobbs and the students he introduced to Greenland would go on to shape American polar science for the next half century or more.

Gould went on to the chairmanship of the geology department of Carleton College, becoming its president in 1945. Like William Carlson, he would later have the school library named for him and, 2 years after his death at the age of 98, the national Science Foundation's ice-strengthened research vessel the R/V *Laurence M. Gould* (see: Carleton.edu, n.d.). And Carlson's friendship with Shorty Cramer would come back to have a profound impact on the American relationship with Greenland (Fig. 8.6).

While the work of William Hobbs and his students accounted for a significant amount of America's scientific knowledge of the west coast of Greenland, it was a privately funded explorer, Louise Arner Boyd, who gave America and the world its first significant imagery and cartography of the east coast. Boyd was the daughter of the part-owner of a California gold mine, and when her two brothers passed away as teenagers and her parents soon followed, Boyd at the age of 33 as the sole heir to a fortune. Her exploring career began in a touring car during a chauffeured road trip across America, at a time when actual roads were few and far between. But it was during a tourist visit to Norway and Svalbard in August of 1924, when she saw polar ice for the first time, that Boyd began the planning of what would become a lifetime of scientific exploration of the North (see Kafarowski, 2017).

Fig. 8.6 Hobbs in his office at the University of Michigan, 1950 (source: William Herbert Hobbs papers, Bentley Historical Library, University of Michigan)

A 1926 charter of the Norwegian ship *Hobby* for a summer of photography in the Franz Josef Land archipelago brought Boyd to the attention of newspapers worldwide. In 1928, she intended to repeat the expedition when the crash of the Italian polar airship *Italia* and the subsequent disappearance of Roald Amundsen during an aeronautical search, led Boyd to redirect *Hobby* on a 10,000-mile search for any survivors. No survivors were found, but Boyd's selfless and self-financed search party that spanned the distance from Franz Josef Land in the east to the coast of East Greenland in the west led to her recognition as the first American woman to receive the Order of Saint Olav, Norway's only order of chivalry at the time (Fig. 8.7).

In 1931, Boyd began the first of her expeditions to the east and northeast coasts of Greenland. Her earlier experiences in Norway led her to charter the 125′ sealer *Veslekari*, launched in 1918 from the same shipyard where Roald Amundsen's *Maud* had been built (see Kjær & Sefland, 2005). Boyd would go on to charter *Veslekari*, under the command of Norwegian ice master Captain Johan Olsen, for all four of her expeditions to East Greenland. Her approach to the island, sailing from Scandinavia, also allowed her to avoid the almost literal boys club of Greenland voyages organized to the west coast by Donald MacMillan and Robert Bartlett (Fig. 8.8).

Fig. 8.7 Louise Arner Boyd in Tromsø, in an image taken by the Norwegian photographer Anders Beer Wilse in 1928 (public domain via *det norske nettbibliotket*)

Boyd's objective in 1931 was nothing less than "to photograph and examine every inlet and sound in the area of Franz Josef and King Oscar Fiords" (Boyd, 1948: 2; Fig. 8.9). The expedition collected several thousand photographic images that, among other discoveries, found "a previously unsuspected connection between Kjerulf and Dickson Fiords" (*ibid.*).

> It aroused in me a desire not only to continue until I had seen all of the region that is ordinarily accessible by ship and to record it with the camera as well and as thoroughly as possible, but also to do what I could to add to the knowledge of features and conditions that everywhere seemed to call for scientific investigation. Also, I gained the familiarity with weather and ice conditions that is necessary if an expedition is to be so planned and equipped that there is good prospect of achieving its objectives during the all-too-brief summer months when the chances of a ship's getting through the coastal ice belt and out again are reasonably favorable (*ibid.*: 3).

As on her subsequent Greenland expeditions, Boyd was strongly supported by the American Geographical Society, whose cartographers either accompanied Boyd or later used her photogrammetry to construct the most detailed charts of the East Greenland coast yet undertaken. The society at the same time was supporting another wealthy American explorer, the Harvard electrophysiologist and polymath Alexander Forbes, who between 1931–35 employed Scottish-American

Fig. 8.8 *Veslekari* in 1928. Source: National Library of Norway (public domain, via Wikimedia Commons)

cartographer O.M. Milller's technique of using oblique aerial images to chart the coastline of northern Labrador from Cape Chidley south for over one hundred miles and inland another 15–20 miles (see Forbes, 1938).

Boyd as well began to carry to East Greenland the enormous, large format aerial survey cameras used by Forbes over Labrador. Without an aircraft of her own, Boyd instead mounted these cameras on tripods which a group of sturdy *Veslekari* crew would haul overland to set them up where Boyd wished to gather her imagery. Boyd also commenced the large-scale collection of Greenland flora, under the tutelage of botanist Alice Eastwood of the California Academy of Science, who analyzed the botanical collections from each of Boyd's expeditions.

As with Forbes, whose Labrador survey vessel *Ramah* he had custom designed and built, Boyd's wealth allowed her the freedom to have extra cabins and storage built on the deck of *Veslekari* for each of her expeditions. These including a separate cabin for herself, another just for her many cameras and film, yet another to serve as a dedicated darkroom, and an entire second galley for her expedition staff, the existing galley being only large enough for the vessel's crew (Fig. 8.10).

By 1933, Boyd's experience allowed her to charter *Veslekari* from Norway to East Greenland staffed with a professional surveyor and assistant surveyor, a physical geographer, a geologist, with herself as photographer and leader of the expedition. The surveying led to three new large-scale maps of the previously unexplored Gregory Valley at the head of Franz Josef Fjord. An echo sounder allowed *Veslekari* to record depth soundings throughout all of the fjords Boyd explored, as well as soundings of the Greenland Sea on both the outward and homeward journeys. "My photographic work added several thousand views to my collection of photographs

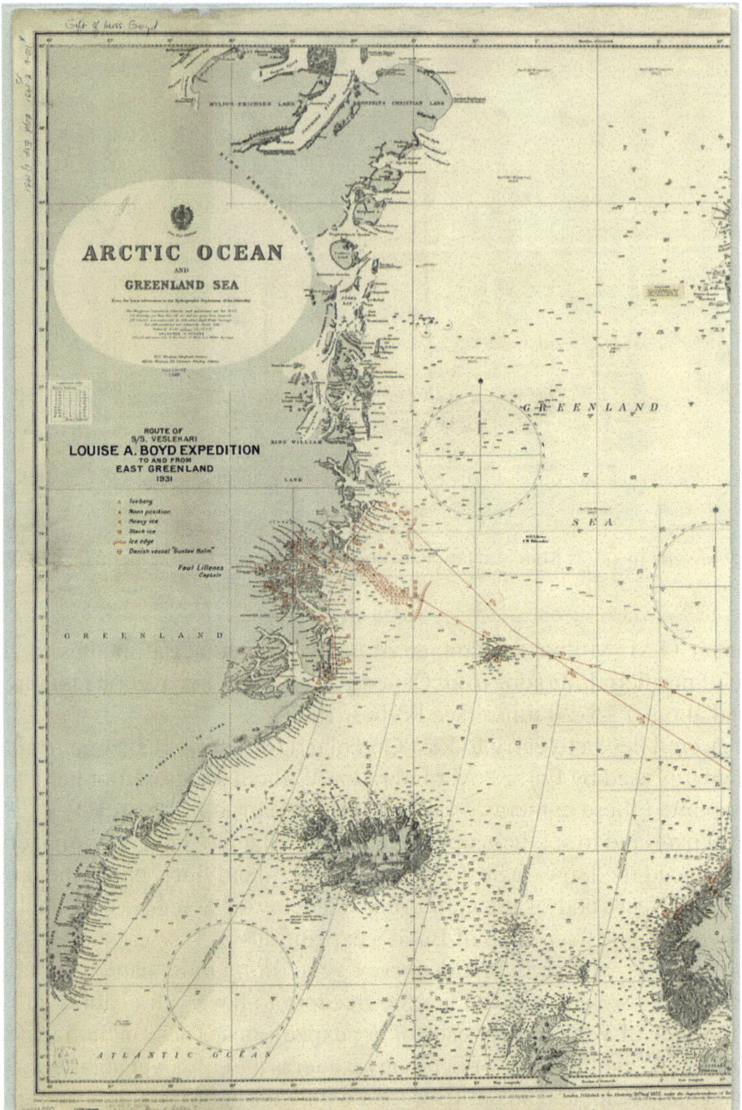

Fig. 8.9 Boyd's approach to East Greenland in 1931. From Arctic Ocean and Greenland Sea, from the latest information in the Hydrographic Department of the Admiralty; Annotated by Louise Boyd in red to show: icebergs, noon positions, heavy ice, slack ice, ice edge, Danish vessel *Gustav Holm*, routes and dates (courtesy of American Geographical Society Library, University of Wisconsin-Milwaukee Libraries)

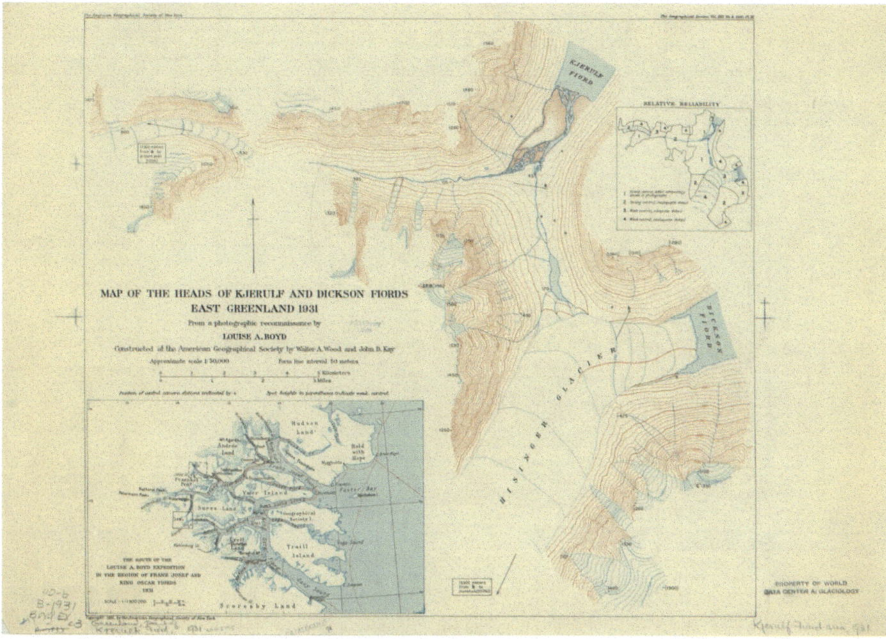

Fig. 8.10 Map of the heads of Kjerulf and Dickson Fiords, East Greenland, 1931/from a photographic reconnaissance by Louise A. Boyd; constructed at the American Geographical Society by Walter A. Wood and John D. Kay (courtesy of American Geographical Society Library, University of Wisconsin-Milwaukee Libraries)

of land and sea ice, glacial marginal features, landforms, and vegetation of East Greenland" (*ibid.*: 4) (Fig. 8.11).

Boyd's crowning achievements were the expeditions of 1937 and 1938, when two large motor dories were carried on board *Veslekari* (for 1938 one of them had in addition its own echo sounder) (Fig. 8.12). In addition to her usual assortment of cameras, Boyd added enormous aerial survey cameras that were mounted on tripods and used for all distant and semi-distant mapping work, including a Fairchild Aero K 6 which had never before been used for anything but aerial photography. The survey work was carried further north from the King Oscar-Franz Josef Fjord areas.

> Since it was felt that a contribution to the studies of glacial recession, which would continue to be an important feature of our geological investigations, might be made through investigation of the major plant communities associated with recessional features, a botanist with special interest in this field was included on the staff in 1937. Greater emphasis was to be put on the hydrographic work than had been the case on any of my previous expeditions and a hydrographer, who was to be in charge of the sounding program, current studies, and tide-gauge recording, was therefore added to the staff… Another major objective was to see how far north we could get along the coast… The attempt was finally made in 1938 and with even better success than we had hoped for (*ibid.*: 4–7).

The results of Boyds 1931 and 1933 Greenland expeditions, spanning the coast from Scoresby Sound to Home Foreland, were published by the American

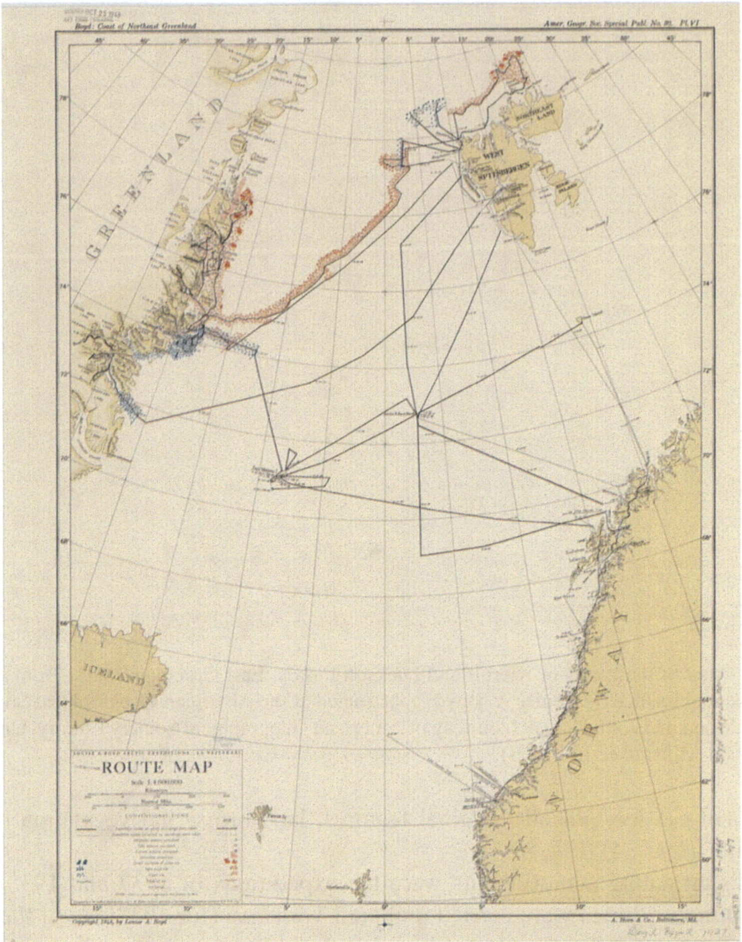

Fig. 8.11 S.S. *Veslekari* hydrographic surveys along the East Greenland coast: Antarctic Harbor to Home Foreland/compiled by J.M. LeRoy, Hydrographer and F.W. Buhler, Surveyor and drawn at the American Geographical Society of New York, 1941; A. Hoen & Co. Coast of northeast Greenland, with hydrographic studies in the Greenland Sea. The Louise A. Boyd Arctic expeditions of 1937 and 1938 (courtesy of American Geographical Society Library, University of Wisconsin-Milwaukee Libraries)

Geographical Society in 1935 (see Boyd, 1935). However, the results from 1937 and 1938, covering the coast 250 miles further north, to Ile de France, would not see the light of day until 1948. U.S. Coast Guard officer Edward H. "Iceberg" Smith, the service's expert on both Greenland and Arctic ice conditions, was stating the obvious when he wrote that the delay in publishing the results of Boyd's surveys further north suggested "its value to our own country.

> Information regarding Northeast Greenland prior to our entering the war was extremely scanty, and in preparing for a mission there during the summer of 1940 I was advised to

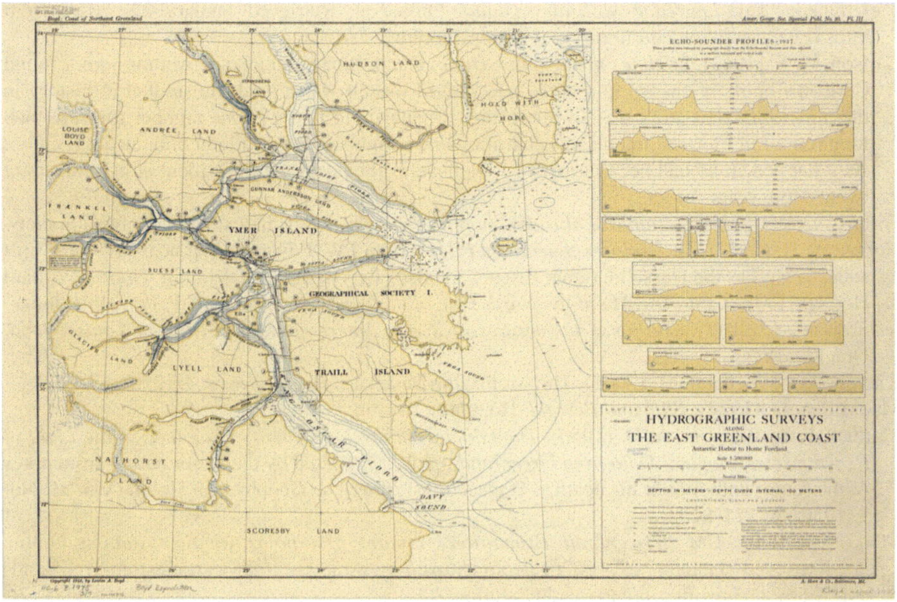

Fig. 8.12 From: *Amer. Geogr. Soc. special publ. no. 30, Pl. VI.* "Coast of northeast Greenland, with hydrographic studies in the Greenland Sea. The Louise A. Boyd Arctic expeditions of 1937 and 1938." (courtesy of American Geographical Society Library, University of Wisconsin-Milwaukee Libraries)

> consult Miss Boyd. The opportunity quite happily arose while the old cutter *Northland* was being hurriedly recommissioned at Oakland, California, prior to proceeding via the east coast of the United States to Greenland (Boyd, 1948: ix).

The service Louise Boyd's self-funded series of expeditions had provided was incalculable as the Second World War loomed. The detailed charting of the East Greenland coast by *Veslekari* under her overall leadership became one of the United States' most closely guarded secrets during the war to come. And, as *Veslekari* arrived back in northern Norway in the fall of 1938, that war was very close indeed. The conquest of Denmark and the occupation of Norway by Germany were mere months away. Greenland suddenly became a strategic weather reporting station, a critical source of materials used in aircraft production, a resupply base for the support of the United Kingdon against Germany and, along the very coast Boyd had surveyed, a potential steppingstone for a Nazi invasion of North America.

References

Bartlett, R. (1940). Greenland from 1898 to now. *National Geographic Magazine, LXXVIII*(1) July.
Boyd, L. A. (1935). *The fiord region of East Greenland.* American Geographical Society.
Boyd, L. A. (1948). *The coast of Northeast Greenland.* American Geographical Society.

Carleton.edu. n.d. *Laurence McKinley Gould*. Online at: Retrieved January 22, 2026, from https://www.carleton.edu/library-exhibitions/gould/
Carlson, W. S. (1962). *Lifelines through the Arctic*. Duell, Sloan and Pearce. Author note: I found Carlson's book by chance in a book sale when I myself was a graduate student in history in the mid-1980s. His story was so interesting I managed to find a phone number for him, then long retired in Florida, and he was charming and insightful during a long phone conversation. As it happened, he had been one of two finalists for the position of president at my university twenty-five years earlier.
Forbes, A. (1938). *Northernmost Labrador, mapped from the air*. American Geographical Society.
Hobbs, W. H. (1952). *An Explorer-Scientist's Pilgrimage*. J.W. Edwards. The near-drowning episode, as well as the Hassell/Cramer rescue, are also covered in some detail in: Hobbs, William Herbert. 1930. *Exploring About the North Pole of the Winds*. New York: G.P. Putnam's Sons.
Kafarowski, J. (2017). *The polar adventures of a rich American dame: A life of Louise Boyd*. Dundurn Press.
Kjær, K.-G., & Sefland, M. (2005). The Arctic ship Veslekari. *Polar Record, 41*(216), 57–65.
MacMillan, D. (1927). *Etah and Beyond*. Houghton Mifflin.
Matthes, F. & Belmont, A. D. (1950). *The Glacial Anticyclone Theory Examined in the Light of Recent Meteorological Data from Greenland*. Part 2. (Prepared by the junior author from notes of the senior author after his death.). Transactions, American Geophysical Union (Vol. 31, No. 2, 1950), p. 174–82.
Mott.org. (1990). *UM "Greenland Expedition" creates first polar weather station*. Online at: Retrieved January 21, 2026, from https://www.mott.org/news/articles/um-greenland-expedition-creates-first-polar-weather-station/
Putnam, D. B. (1926). *David goes to Greenland*. G.P. Putnam's Sons.
utoldeo.edu. (n.d.) *The Heart of Any University: 50 Years of the William S. Carlson Library*. Online at: Retrieved January 22, 2026, from https://www.utoledo.edu/library/virtualexhibitions/CL50VX/exchap2.html

Chapter 9
Masters of the Arctic: "Iceberg" Smith, Bernt Balchen, and the War Below Zero

Abstract The work of William Hobbs and Louise Arner Boyd in the 1930s gave the U.S. military the beginnings of the data required to start the defense of Greenland even before the formal American entry into the Second World War. The final pieces of that puzzle were added by two of the most remarkable Arctic hands in American history: the U.S. Coast Guard officer Edward 'Iceberg' Smith, and the Norwegian-American aviator Bernt Balchen.

Keywords Greenland · Greenlandic Inuit · Isolationism · U.S. Coast Guard · North Pole · Second World War · Greenland Patrol · Marion expedition · Cryolite · Ivigtut · Bluie · Narsarssuak · Operation Bolero · Søndre Strømfjorden · Ice Cap

> *When you fight in the Arctic, you fight on the Arctic's terms.*
> —Bernt Balchen, *War Below Zero: the Battle for Greenland* (1945)

> *Then against the horizon I saw a series of warts in silhouette. They were sharp-edged and I supposed them to be optical illusions, a final adventure in my private trance. But they persisted like the stars and slowly enlarged until I knew they must be real. There, then, was the enormous mass of Greenland, and I could not look away from it.*
> —Ernest K. Gann, *Fate is the Hunter* (1961)

William Hobbs' theory of a permanent glacial anticyclone over the Greenland ice came in for much criticism during his lifetime. But the rescue of Fish Hassell and Shorty Cramer by the Greenlanders visiting the University of Michigan research station in 1928 would have lasting effects far out of proportion than even the international news they created at the time. Hobbs' experience in actually building a runway for land planes in western Greenland, along with the German fluency that allowed him to keep a close watch on German activities in Greenland like the expeditions of Alfred Wegener, made him an important voice on the subject of aviation and Greenland and their intersection in the defense of the Western Hemisphere. These in turn would become almost as critical to the American entry into the Second World War as the attack on Pearl Harbor itself.

P. J. Capelotti, *Lost Greenland*, Springer Polar Sciences,
https://doi.org/10.1007/978-3-032-27841-8_9

In 1940, Hobbs would take his place as an arch foe of Charles Lindbergh as the Lone Eagle descended into his flirtation with isolationism, antisemitism, and Nazism. Hobbs joined a group called the "Committee to Defend America by Aiding the Allies," where his main role was "advising the government concerning the defense of Greenland" (Hobbs, 1952: 157).

> In 1940 (Lindbergh) began to claim in radio broadcasts that we should not attempt to defend Greenland, which President Roosevelt had pointed out would give Germany a springboard to use against us if she could secure a foothold there. Between 1929 and 1932 she had had the great Wegener expedition there… I had pointed out Greenland's vital importance in our defense and the danger of letting Germany gain a foothold there… Lindbergh had particularly urged that the climate, as well as the rugged topography of the Greenland coastline, would make flying there impractical (ibid.: 157–159).

Hobbs recommended that a survey flight be sent to Greenland's west coast to search for other potential airfield sites than his makeshift runway near Mount Evans created by his students in 1928. The survey flight, led by U.S. Army Air Corps Major Julius Kahn Lacey, who was in the middle of forming the corps' own weather service, recommended expanding Hobbs 1928 site. This would become Bluie West 8, a main aviation and resupply hub for the Allies in Greenland during the Second World War. Hobbs in addition advised, from his brief visit to the area some years earlier, that a more southerly airfield be considered near Julianehåb (Qaqortoq) and, just a few miles from the spot Hobbs recommended, Bluie West 1 was constructed, and served as an almost exact midpoint for aircraft flying from Labrador to Iceland.

One American who echoed Hobbs' insistence on bases for landplanes was one of his former students, William Carlson, by the late 1930s a professor of education at the University of Minnesota. Carlson's friendship with the doomed Shorty Cramer, along with his leadership of one of the University of Michigan Greenland expeditions, had left a deep impression on him, so much so that Professor Carlson joined the Army Air Corps as an expert on both Greenland and aviation in the Arctic.

> "After Cramer had pioneered it, there was continuing activity over Greenland where the icecap was the principal obstacle to the planes of that period. Wolfgang von Gronau crossed the icecap north of Scoresby Sound three times as he flew between Germany and Canada. Farther south, the British Arctic Air Route expedition (under Gino Watkins) used a small amphibian plane for crossing…
>
> "However, when the northern airway was needed a few years later, Lindbergh's lack of enthusiasm was smothered by two loud and clear voices shouting the possibilities of aviation in the Arctic: Professor William H. Hobbs, the great teacher and Arctic scientist, and the expert aviator and explorer of the north, Bernt Balchen. They repeatedly called attention to Greenland in Balchen's terms, as the "great aircraft carrier of the Arctic." This was during the period when long-range German aircraft were making reconnaissance flights over the icecap.
>
> Hobbs and Balchen deserve the principal credit for American ventures in Greenland. Instead of becoming a Nazi jumping-off place in the Western Hemisphere, it became a major American bastion in the struggle to whip Hitler (Carlson, 1962: 49–50).

Bernt Balchen was born in southern Norway near Kristiansand in 1899. His mother left the family when Balchen was four and thereafter his father, the local doctor, took it upon himself to raise the boy, putting him on ski at six and teaching him to race and jump (Glines, 1999: 16–19). Balchen became not only an expert skier but a champion boxer and wrestler, as well as a crack shot with rifle and pistol, hunting only to feed himself and never for trophies. He brought a sketch pad into the forests where he skied and became a creditable self-taught artist. He learned English, French, and German and, through a great-uncle, was introduced to Norwegian national icons Fridtjof Nansen and Roald Amundsen. When the teen-aged Balchen told Amundsen he wanted to become an explorer, Amundsen advised him to "keep on skiing and camping and get in real good trim, besides getting good marks in school… You must learn about previous expeditions and benefit from their experiences, and be thorough and well-prepared in all respects. Above all, you must learn to take care of yourself, to obey and give orders and to work with others" (*ibid*: 19).

In the last year of the First World War, Balchen joined the French Foreign Legion but before he could see action Norway called him home to serve in the army. When Russia invaded Finland, Balchen was sent there and participated in a cavalry charge against a Russian trench that nearly killed him. Recuperating at home, he resigned his army position and entered the Norwegian Naval Air Force pilot training program in 1920 and received his wings the following year. In 1925, when Amundsen went missing for several weeks over the Arctic Ocean, Balchen was selected to make search flights for him. As he arrived in Svalbard, Amundsen and his crew made their miraculous return to Kings Bay. "I remember you," said Amundsen as he and his men came ashore. "You told me you wanted to be a great discoverer and go on an expedition with me. You see, I don't forget" (*ibid.*: 25).

The following year, true to his dream, Balchen served as a member of the Norwegian ground crew for Amundsen's flight to the North Pole in the airship Norge. It was the same summer, and the same place, where Balchen would meet Richard Byrd for the first time, a meeting would profoundly impact the remainder of Balchen's life. Byrd's limited experience with aircraft during Donald MacMillan's 1925 Greenland expedition, combined with Amundsen's near approach to the pole that same summer, had convinced Byrd that he could earn eternal fame with a quick flight to the pole and back.

When Byrd's ski-equipped plane cracked both ski and a support strut during an attempted take-off, Balchen used his expertise to remove and repair the ski and strut and then burned a combination of resin and pine tar into the ski to enable them to slip off the ice even under the heavily loaded aircraft. Byrd and his pilot Floyd Bennett took off towards the North Pole and returned just 15 h and 30 min later, with Byrd immediately sending a radio message to the United States that he and Bennett had become the first aviators over the pole. It would be half a century before multiple historians and meteorologists concluded that Byrd's flight would have been impossible in the time he claimed. It would be another quarter century before Byrd's own logbook was examined and seemed to confirm the fraud (Wilford, 1996).

Byrd invited Balchen to America to join his flight team for another Greenland flight the following year, but instead 1927 became the great air race to Paris eventually won by Lindbergh. Since Balchen was the only one of Byrd's pilots who could fly on instruments, and with his great friend and colleague Floyd Bennett ill, he became chief pilot of Byrd's trans-Atlantic flight conducted in the wake of Lindbergh's triumph. Byrd then turned his ambition to Antarctica, and with Balchen as his pilot, claimed to have reached the South Pole in late November 1929. Once again, Byrd made his claim the absence of any actual navigation on his part. "As on the flight across the Atlantic, Byrd had not taken a single sextant shot to ascertain that they had actually reached the South Pole.

There are no significant landmarks that identify the vicinity of the South Pole, so it was determined solely by Balchen computing the speed and the time that the aircraft had flown and estimating when the aircraft *should* have arrived over the southernmost point on earth" (Glines *op cit*.: 90) (Fig. 9.1).

Balchen became an American citizen and a pioneer of airlines in the north, including one of the forerunners of Scandinavian Airlines. When Russian again invaded Finland, this time in 1939, Balchen was negotiating the sale of aircraft from the U.S. to Finland. Global war was coming on and fast. It was only a few months later that Balchen's home country came under attack. Germany invaded both Denmark and Norway on April 9, 1940. The Danes capitulated soon after; the Norwegians held out until June. The Danish North Atlantic colony of Greenland thereafter loomed as a potential advance base from which Hitler could stage an invasion of North America.

With lines of communication severed between Copenhagen and Danish colonial offices in southwest Greenland, and Oslo and Norwegian weather stations in northeast Greenland, the Danish governor in Greenland, Eske Brun, recast the island as independent of occupied Denmark. In May, Brun asked the United States for protection, and such an agreement was formalized the following April. The U.S. Coast Guard would provide the protection, and in return the U.S. would receive the cryolite from Greenland.

Although the Greenland Patrol was formally organized on October 25, 1941, the U.S. Coast Guard had long experience in both Greenland and Arctic waters prior to formal American entry into the Second World War later that year. This experience included more than a half century of Bering Sea patrols in Alaskan waters and a quarter century of International Ice Patrols following the sinking of the Titanic in 1912. This body of experience had created a cadre of Coast Guard officers who were expert navigators in Arctic conditions.

Such experience was called upon almost 2 years before the formal entry of the United States into the Second World War, and nowhere so heavily as in the person of Edward H. "Iceberg" Smith. Smith's scientific expedition to the Davis Straight with the cutter *Marion* in 1928 laid the basis for much of the service's knowledge of the environmental conditions around Greenland. Smith earned his nickname even before he advised Fish Hassell in northernmost Labrador in 1925.

Beginning in 1921, as a young Lieutenant, junior grade, Smith was tasked with writing the annual report of the International Ice Patrol. Born on the island of Martha's Vineyard off Massachusetts in 1889, Smith descended from a family of whalers. After a year at the Massachusetts Institute of Technology, he entered the

Fig. 9.1 Floyd Bennett and Bernt Balchen in 1929 (public domain via the Bernt Balchen Collection, San Diego Air and Space Museum Archive)

Coast Guard Academy in New London, Connecticut, when the academy was still known as the Revenue Cutter Service School of Instruction and classes were held aboard the Revenue Practice Cutter *Itasca* at Arundel Cove, Maryland (Capelotti, 1996).

In September of 1919, Smith was assigned as navigator and scientific observer to the *Seneca*, which left for Ice Patrol duty in the spring of 1920. His presence marked the first time a Coast Guard officer was detailed as scientific observer during a Coast

Guard cruise, conducting experiments and observations in furtherance of oceanographic knowledge. Smith made the most of this opportunity, transferring from ship to ship so that he could remain at sea conducting observations from the time the patrol began in February until its conclusion in July. From then onwards, the bulk of Smith's career, from 1920 until his retirement as rear admiral in 1950, was spent studying the oceanography of the Arctic and the North Atlantic.

By the early 1920s it became apparent that Coast Guard officers with professional oceanographic training were essential for Ice Patrol duties. To meet this need, the Coast Guard in 1923 established an oceanographic unit at Harvard University charged with providing support and conducting research for the patrol. From 1923 to 1931, the unit consisted of one civilian and one military oceanographer, plus five enlisted oceanographic technicians. The military oceanographer was Smith who, as a result of his annual ice reports, was rapidly becoming known to his fellow officers as "Iceberg." He could look at ice for hours. At meals, it was even said that he would stir the ice in a pitcher of ice water and study the melting ice and the eddies created.

The importance of Smith's research to the safety of life at sea was soon recognized abroad by Norwegian shipping concerns, and an oceanography fellowship from the American Scandinavian Foundation followed. The Coast Guard granted Smith leave from 1924 to 1925 to study at the Geophysical Institute at Bergen, Norway, under the Norwegian oceanographer Vilhelm Bjerknes, the founder of modern meteorological prediction: the so-called Bergen School of Meteorology.

Upon his return to the U.S., Smith standardized the investigations of the Ice Patrol under his own "practical method for determining ocean currents," which was used to track the movements of icebergs. He also initiated a method of iceberg forecasting whereby the number of bergs annually drifting south of Newfoundland were predicted.

Between 1928 and 1936, Smith was nominally the commanding officer of various vessels of the destroyer force the Coast Guard operating to suppress alcohol smuggling, as well as commander of the Coast Guard base at Woods Hole, Mass. Still he was able to maintain his career specialization even as resources were drawn away to enforce Prohibition. In the summer of 1928, Smith was placed in command of the Coast Guard cutter *Marion*, a 125-foot vessel built for offshore patrol duty and roughly the same dimensions as another famous military vessel converted to oceanographic research, Jacques-Yves Cousteau's *Calypso* (Fig. 9.2).

Sixteen years of data from the Ice Patrol had given scientists a good picture of drifting ice as it moved into the Gulf Stream. Smith was now given the opportunity of a lifetime: to take a vessel into the waters west of Greenland, the birthplace of the icebergs, and apply to the northern glaciers and sea ice the same surveying methods he had perfected for the southern bergs. The *Marion* Expedition would become the most comprehensive oceanographic expedition ever conducted by the United States to that moment.

After leaving New London and calling at Smith's hometown of Vineyard Haven, Mass., the *Marion* departed Boston for Davis Strait and Baffin Bay on July 11, 1928. Little more than 2 months later, the 73-day cruise to the Arctic ended after covering 8100 miles, during which Smith and his team of oceanographers surveyed

Fig. 9.2 The crew of the *Marion* in New London, CT., in June 1928, just prior to the Greenland expedition. LCDR Edward H. "Iceberg" Smith, commanding officer, stands beaming just to the left of the upper ladder (public Domain via U.S. Coast Guard Historians Office)

an area of nearly half a million square miles between Greenland and Canada. All this cruising evidently took the land legs from the *Marion* crew, for when native Greenlanders challenged the Coasties to a soccer match at Godhavn (Qeqertarsuaq), the Americans lost 28-0.

The *Marion's* crew was more successful with its science. They made more than 1900 recordings of water temperature and salinity at 190 observation stations. They employed Greene-Bigelow water bottles lowered on 3 miles of steel cable. A bottom sampling tool collected mud and ooze from the ocean floor. Combined with measurements made by a Danish expedition aboard the vessel *Godthåb*, the *Marion* expedition was able, in Smith's words, "to present a comprehensive exposition of Arctic ice, its state, behavior, and distribution to the western North Atlantic" (*ibid.*) (Fig. 9.3).

A year and a half later, Harvard University awarded Smith a Doctor of Philosophy in geological and oceanographic physics. At the same moment, the Woods Hole Oceanographic Institution was established and a year later the Coast Guard moved its oceanographic unit from Harvard to this small seaside village of ocean scientists in the southwest corner of Cape Cod.

The Coast Guard added a third 125-foot cutter to the Ice Patrol in 1931. The *General Greene's* sole mission was oceanographic research, and it allowed Smith to gather his data without the distraction of tracking icebergs in the shipping lanes. In the same year, Smith was appointed as scientific crew of a proposed north polar flight of the German dirigible *Graf Zeppelin*. The 8000-mile flight took place from July 24 to August 1, 1931, the longest flight the airship ever made. The *Graf Zeppelin* did not reach the pole, but it did survey vast sections of the Franz Josef Land archipelago, with Smith serving as navigator and science officer.

By 1940, Smith was commanding officer of the Coast Guard cutter *Northland* as it commenced what was scheduled to be a three-year survey of ice conditions around Greenland. The German invasion of Denmark and Norway scrubbed those plans.

Fig. 9.3 The USCG cutter *Marion* off Greenland in 1928 (public domain via the U.S. Coast Guard Historians Office)

Instead, throughout 1940 and 1941, *Northland* led a number of Coast Guard cutters equipped with aircraft as they began to cruise both the east and west coasts of Greenland in search of sites for naval, air, and weather stations, and to create new charts to replace the former Danish charts now in occupied Copenhagen.

In northeast Greenland, *Northland* liaised with the free Norwegian navy inspection vessel *Fridtjof Nansen*, under the command of Norwegian Ernst Ullring, itself in the process of destroying Norwegian radio stations and removing Norwegians suspected of collaboration with the Nazis. At Ivigtut in October, Smith drew up a defensive scheme for the Arsuk Fjord, site of the cryolite mine, and sited a 3″ gun there that had been delivered earlier by the cutter *Campbell* (Capelotti, 2005: xiv-xix).

Smith's many and various activities—charting, defense of undefended Greenland, protection of the cryolite mine at Ivigtut, denial of the Greenland fjords to German submarines, weather stations, and patrol aircraft, blocking Nazi ambitions in the western hemisphere—was one of the most closely-guarded secrets of the war. By the end of 1941, all of these activities were gathered together and designated as Task Force 24.8 and attached to the Commander in Chief of Atlantic naval forces. Most called it the Greenland Patrol. Samuel Eliot Morison described it thusly: "The commander, "Iceberg" Smith in cutter Northland, received orders from the Chief of Naval Operations to do a little of everything—the Coast Guard is used to that" (Morison, 1947: 62–63).

The cryolite was mined from deposits located at Ivigtut on the island's southwest coast. This quartz-like mineral, used in the electrolytic production of aluminum, was essential for construction of aircraft. The mine's production quantity dwarfed the total production of the only two other known mines in the world, one in Colorado and the other in the Ural Mountains of the Soviet Union. Greenland cryolite had been a cornerstone of the U.S. aircraft industry since the 1920s and accounted for practically all of Greenland's exports. It would become an essential element in the production of more than a quarter million combat aircraft in the U.S. and Canada during the war.

The Ivigtut mines are less than a mile up the Arsuk Fjord from the sea, and vulnerable to enemy attack. Vessels transporting cryolite from Greenland to the U.S. were also in danger, particularly when moving southbound through the narrow Strait of Belle Isle. In May 1940, an American Consul—the first in Greenland—was transported to Ivigtut by the U.S. Coast Guard cutter *Comanche*, then delivered further up the west coast to Godthåb, where a rudimentary American consulate was established. To safeguard Greenland's (and America's) neutrality, ex-Coast Guard sailors were sent to Ivigtut in August, armed to defend the mines. In addition to its cryolite, Greenland was strategically desirable in several other ways. It provided a refueling steppingstone for short-range, England-bound U.S. military aircraft, and served as weather observation outpost for North Atlantic convoys.

In early 1941, the U.S. passed the Lend-Lease Act and, defying German sea blockades, began supplying the British with large amounts of munitions and other war materials. These shipments transited Greenland waters. At the same time, a South Greenland Survey Expedition on board the U.S. Coast Guard cutter *Cayuga* searched for appropriate locations to site airfields, seaplane bases, aids to navigation, and radio and weather stations. In April, the U.S. formalized an agreement with the Danish ambassador, Henrik de Kauffman—who claimed the authority of the Danish crown with the Danish government under German control—that placed Greenland under the protective custody of the United States. Once President Franklin D. Roosevelt approved the treaty in June, the United States received, for all intents and purposes, unlimited military access to the island. The U.S. extended their forces to an occupation of Iceland on July 1.

For Roosevelt, as for Churchill, the North Atlantic sea lines of communication were an obsession, and Greenland was the fulcrum around which those supply lines turned. The U.S. Army used the data gathered by the South Greenland Survey Expedition, along with the recommendations of Hobbs and Balchen, to move rapidly to develop Greenland as an 'American aircraft carrier. 'Thirteen sites were proposed as potential bases and code-named 'BLUIE.' The sites ran along the west coast from Narsarssuak north to Søndre Strømfjorden near Professor Hobbs' base, and on the east coast from Kap Farvel north to Eskimonæs on Clavering Island at 74 degrees north near Lauge Koch's research station. Samuel Morison's official history of U.S. naval operations during the war was likely the first to be allowed to publish a list of the top secret stations, compiled by a navy yeoman, Harry E. Foster (*ibid.*: 62).

Blue East 1, Torgilsbu, the base near Cape Farewell
Blue East 2, Ikateq on Angmagsalik Island, opposite Iceland, radio and meteorological station*
Blue East 3, Gurreholm on Scoresby Sound
Blue East 4, Ella Island on Davy Sound, at lat. 73°N
Blue East 5, Eskimonæs, at lat. 74°N, radio (seized by Germans 23 Mar. 1943, and when reestablished, located at Myggbukta)
Bluie West 1, Narsarssuak, the main U.S. Army and Navy base and airfield
Bluie West 2, Kipisako
Bluie West 3, Simiutak, at mouth of Skovfjord, a HF/DF station
Bluie West 4, Færingerhavn, at lat. 63° 40'N
Bluie West 5, Godhavn, on South Disko Island
Bluie West 6, Thule, at lat. 76° 30' N
Bluie West 7, Grondal, protection of cryolite mine at Ivigtut
Bluie West 8, Søndre Strømfjord, emergency landing field, lat. 67° N
(*Author's note: This refers to a station near Angmagssalik itself. The airfield thirty miles north was also referred to as Bluie East 2.)

The most promising of the bases, by far, was Bluie West 1, near the village of Narsarssuak in southwest Greenland, and located in the same area identified as similarly promising nearly a thousand years earlier by Eirik Raude, the Norse explorer Eirik the Red. Like Eirik, there was another newcomer to Greenland looking to hide from debts and the pursuit of the law, one Elliott Roosevelt, the son of the president. Elliott Roosevelt participated in long-range survey flights from Canada to Iceland and recommended, among other sites, the strip of shore north of Angmagssalik called Ikateq that would become Bluie East 2 (Hansen, 2012: 176–187).

In his memoirs, Henry Harley "Hap" Arnold, a military aviation pioneer who had been taught to fly by the Wright Brothers aviation school in Ohio and by 1941 had become Chief of the Army Air Forces, overstates Elliott Roosevelt's role in the survey of Greenland. But what could not be overstated was the benefit of having the president's son in your corner as you asked Congress for an increase in the U.S. Army Air Corps to over two million personnel, 60,000 aircraft, and landing fields and support bases around the globe, a massive program of armament and reach that in 1942 was increased to nearly three million personnel and 75,000 aircraft.

> Elliott Roosevelt was the first man to come to me with a plan for flying over the ice cap. He had selected a crew, had a plane picked out, and all he needed was my authority.... In many places it would be possible to land airplanes successfully, Elliott thought, so we might even get airplanes in and out in case of emergency. That picture tore away the veil of mystery in which the Greenland ice cap had been shrouded for centuries (quoted in *ibid.*: 187).

Roosevelt's father pushed the development of Greenland bases right along with his son. In an address to the nation in late May, 1941, Franklin Roosevelt warned the country that almost all of the supplies that the U.S. provided to England "go by a northerly route, which comes close to Greenland and the nearly island of Iceland... Nazi occupation of Iceland or bases in Greenland would bring the war close to our continental shores, because they are stepping stones to Labrador, Newfoundland, Nova Scotia, and the Northern U.S., including the great industrial centers of the north, east, and Middle West" (quoted in *ibid.*: 191).

For Hap Arnold, to secure Greenland as an airbase required one aviator more than any other and so, as Roosevelt rallied the nation, Arnold called Bernt Balchen into his office in Washington, D.C. By 1941, Balchen had long been considered the aviator with more experience in flying over ice and snow than any other. Six months before the Japanese attack on Pearl Harbor, Arnold assigned Balchen the secret task to build the series of proposed air bases in Greenland and then command one of them for the duration of the war. Throughout the summer of 1941, Army freighters escorted by U.S. Coast Guard vessels carried enough soldiers and supplies to Narsarssuak to construct eighty-five buildings, three miles of roads, and, eventually, one of the most important air bases of the Second World War, Bluie West 2, and directly across the fjord from where Eirik had sited his farm a thousand years before. To the north, Balchen and a crew of thirty developed a site just ten miles from Hobbs' old camp on the Søndre Strømfjorden. This became Bluie West 8, and by the spring of 1942 the airfield was ready along with a road connected it to the old University of Michigan research station. A smaller field (Teague Field) was developed halfway between Bluie West 1 and 8 and, along with Bluie East 2 in East Greenland, they became the major hubs of Allied activity in Greenland during the war.

> It was a secret war, waged in semi-darkness north of the Arctic Circle, on a remote battlefield perpetually locked under ten thousand feet of solid ice. The weapons were not tommy-guns and tanks; the real heroes of this war were nameless enlisted men working in Air Force ground crews at fifty below zero, or standing on Coast Guard cutters fighting through the

pack-ice, or living all winter long in isolated weather stations along the Ice Cap, buried under eighteen feet of snow...

The problem of constructing a base in such an unknown and virtually uninhabited land was a formidable one. There was not even an Eskimo settlement within a hundred miles of the region where we were going, no available equipment or supplies short of the States. Every last toothpick had to be planned and purchased in advance; here in the Arctic you have to take it with you. We had little or no knowledge of what problems we would face in erecting docks, how close we could get to the site of the landing field before unloading our construction gear, how many miles of road would have to be built. All these matters had to be decided after our arrival, and by then we would have no time to order additional equipment: all water transportation ends with the advent of winter. We had to take along general construction machinery that would meet any emergency (Balchen, 1945: 4; 8).

When the barrel-chested officer with the Norwegian accent (Balchen by then had been commissioned into the U.S. Army Air Corps) would show up at supply depot with requests for enormous amounts of equipment for purposes he could not talk about, he was often met with instant refusal. A single phone call to Arnold's office swiftly swept these obstructions away (Fig. 9.4).

On June 14, 1941, President Roosevelt ordered the American-held assets of Germany and Italy, and the countries occupied by them to be frozen. Less than a month later, U.S. troops were moved into Iceland to shield it and Greenland from

Fig. 9.4 A snapshot of the difficulties in establishing any base in Greenland: ice, snow, mud, stony shorelines, to add to a complete lack of port infrastructure. Here, a landing craft puts a crane ashore as a PBY Catalina flying boat sits wrecked on the shore. Perhaps the only advantage to working in Greenland was that the plywood landing craft did not have to face active opposition as they moved men and materiel ashore (public domain via the U.S. Coast Guard Historians Office)

possible German invasion, and to safeguard the seaways between Iceland, Greenland and the U.S. The U.S. then closed all German and Italian consulates in the United States.

On September 4, a German U-boat launched its torpedoes against a U.S. Navy destroyer, the *Greer*, as it passed Greenland bound for Iceland with mail for the new American base there. Less than a week later, Roosevelt issued his "shoot-on-sight" order, declaring that German or Italian war vessels that entered waters deemed necessary for the defense of the Americas, did so at their peril. One day later, *Northland* seized a sealer flying the Norwegian flag, the *Buskoe*, which was carrying a German radio station to northeast Greenland for the Nazis. The Coast Guardsmen from the *Northland* destroyed the radio and captured the *Buskoe*, the first American naval capture of the war.

On December 11, 1941, four days after the Japanese attack on Pearl Harbor, Germany and Italy declared war on the U.S., and the U.S. Congress quickly responded in kind. The Greenland Patrol under the command of 'Iceberg' Smith was now faced with the daunting task of ensuring the safe transport of Allied shipping to and from Greenland, sea traffic that had suddenly increased by a factor of three.

Smith was suddenly desperate for escort vessels to convoy the increase in Greenland-bound shipping, and to supply the outposts and weather stations of the Greenland coast. Most U.S. weather observation outposts in Greenland were now manned by small numbers of U.S. Army meteorologists, and mail to these outposts could only be delivered by ship during the brief Greenland summer. Otherwise, only if they were fortunate and weather conditions were not too fierce or dangerous, could aircraft parachute in additional mail or supplies. The army meteorologists were a special breed, able to tolerate isolation and privation, with their radio transmitters and receivers as their only lifelines. These outposts were responsible not only for providing meteorological data, but also to assure that no enemy presence would function in their vicinity for similar purposes.

So when Commander Smith located ten 120-foot fishing trawlers in Boston, he immediately recognized their strategic potential, and cabled Vice Admiral Russell R. Waesche, Commandant of the Coast Guard. Waesche commandeered the trawlers for the Greenland Patrol and manned the vessels with hand-picked crews.

Both Waesche and Smith knew this to be but a temporary expedient, but more appropriate vessels would not be ready for more than a year. Until then, this small fleet of ten vessels had to fill the yawning breech in American naval strategy that stretched from Boston to Narsarssuak, and supply remote outposts flung along the length of a hostile and forbidding shore. The ten vessels converted into expedient Arctic cutters were all in the 120-foot, 225-ton class. They were all assigned Greenlandic names: *Aivik*, *Aklak*, *Alatok*, *Amarok*, *Arluk*, *Arvek*, *Atak*, *Natsek*, *Nanok*, and *Nogak* (Fig. 9.5).

As Iceberg Smith wrote later:

> "Greenland's needs were immediate when the Coast Guard was called upon to make its first visit there. All supplies which had formerly come from Denmark, had to be purchased in the United States, and the shipping and delivery had to be completed during open navigation…

Fig. 9.5 The doomed converted trawler *Natsek* (WYP-170) of the U.S. Coast Guard's Greenland Patrol, seen leaving Boston in June 1942 on the cruise from which it would it not return. It was lost in the Strait of Belle Isle in December of that year with the loss of all hands (public domain via the U.S. Coast Guard Historians Office)

> "One of the first tasks curiously enough was an Army job. The only commercial cryolite mine in the world, located in Ivigtut, Greenland, required immediate protection against the possibilities of enemy attack ... and the crew of the cutter Northland after herculean efforts succeeded in mounting a 3-inch gun on a mountainside commanding seaward approach to the fjord...
>
> Coast Guard cutters ran lines of soundings into the principal fjords to provide early cargo and troop ships with a semblance of piloting safeguards.... The fact that several of the cutters cruised along the Greenland coast, assisted to familiarize them with the country which the United States had been called upon to defend (Smith, 1949: 7)

As if this were not enough, Smith knew he needed eyes and ears on the remote coastline of northeast Greenland, the most likely area for the Germans to attempt operations. There, with the enthusiastic cooperation of Greenland's governor, Eske Brun, Smith created a "small group of brave Danish, Norwegian, and Greenlandic (Inuit) hunters ... organized to patrol a stark 700-mile stretch of Greenland's coast by dogsled in search of German interlopers. Operating hundreds of miles above the Arctic Circle, the North-East Greenland Sledge Patrol was one of the most unique military units of the war" (Desh, 2023).

Along with Hap Arnold's Operation Bolero, the aerial ferry operation to deliver fighters and bombers to England, the ten commandeered trawlers of the Greenland Patrol would supply Greenland until proper port facilities could be constructed around the island. One of the first to arrive on this new supply route was William Carlson, who landed at Bluie West One on the afternoon of April 12, 1942. It was his first visit to Greenland since the end of the University of Michigan expeditions.

> We were pioneering the northern route to the United Kingdom; the Liberator that probed toward the ice-encrusted coast ahead was the first four-engined plane to essay so hazardous a mission. We knew something of the ferrying operation that was planned if we succeeded; it is unlikely, however, that any of us dreamed, during that first flight, of the thousands upon thousands of Fortresses, Liberators, Marauders, and Mitchells, which would roar through

> the Arctic air bound for Great Britain and the battle for the skies of Europe (Carlson *op cit.*: 4).

That route would encounter innumerable obstacles, solved by the ingenuity of Arctic veterans like Smith and Balchen. Balchen may have been appointed base commander at Søndre Strømfjorden, but he was anything but a desk jockey. As Carlson wrote of their time in Greenland: "It isn't likely that any of the veterans of the winter of 1942–42 along the northern route looks back upon those months with nostalgia. There aren't many men who thrive on hardship and discomfort. Perhaps Bernt Balchen is one; the tougher the job he was offered, the more his round Norse face would blossom" (*ibid.*: 74).

Just 2 months after Carlson's arrival in Greenland at the start of Bolero, a B-17 Flying Fortress named *My Gal Sal* was forced to make an emergency landing on the ice cap. Pilot Lt. Ralf Stinson landed the huge aircraft wheels up and slid to a safe stop with only some damage to the propeller blades. Balchen, by now a Colonel in the U.S. Army Air Corps, set off in a PBY Catalina and found a lake to land on some 25 miles from the B-17. Balchen and two Army sergeants, Hendrick Dolleman and Joseph Healey, and a team of dogs, set off for the downed bomber. Both Healey and Dolleman had extensive experience in polar travel in Antarctica.

> You think of the Greenland Ice Cap as a solid substance, but actually this is not the case. Ice in the Arctic is fluid, constantly moving and shifting as it works towards the sea; and near the coast, where the glaciers break off into the ocean, the bending produces deep vertical cracks in the surface of the Cap. Some of these crevasses are virtually bottomless: you can drop an object into one of them and listen in vain for it to land. Most of the fissures are covered by thin roofs of snow, called snow-bridges, which vary in width and thickness. Sometimes the ice will shift, and one side of a bridge will break loose, forming a natural booby trap in the trail: the only safe procedure is to test ahead of you with your ski-pole every step of the way (Balchen *op cit.*: 16–17).

It took 10 days for Balchen and his team to reach and rescue the downed air crew. "Our rescue had been none too soon," Balchen later wrote. "Two days later I happened to fly over the same area. In vain I looked for the lake on which we hand landed. A large fissure had opened in the ice directly under it; all the water had poured down through, the entire lake had disappeared overnight..." (*ibid.*: 19) (Fig. 9.6).

The *My Gal Sal* rescue was not the most difficult Balchen undertook even in 1942. During that year, in aid of their survival in Greenland, he taught all of his rescue flight crew how to ski and to identify edible plants along with birds and fish. When the commander of Bluie West 1 went missing in his AT-6 two-seat trainer aircraft, it was Balchen who made an expert lake landing 10 miles away and hiked to the rescue.

But by far the most complex operation took place in November, during the same incident that led to the death of Max Demorest. On November 6, 1842, another B-17 Flying Fortress was being ferried to Europe when it was diverted to search for a missing transport plane in Greenland. The B-17 was piloted by a twenty-seven-year-old from California, Armand L. Monteverde, and his crew consisted of men from Wisconsin, Texas, New York, Pennsylvania, and Kansas. For a second search

Fig. 9.6 Bernt Balchen belly-lands his PBY Catalina flying boat in the Greenland Ice Cap, 1943 (public Domain US Army photo No. C-55800)

flight the following day and a third on November 9, Monteverde's crew was augmented by two sergeants including another Texan and a Kentuckian. Flying low in white-out conditions, Monteverde banked the huge aircraft towards the coast when a wingtip caught the ground, and the B-17 broke in half and skidded to a stop. Two crew members were ejected through rents in the airframe and lay on the surface with broken bones and lacerations. The rest of the crew was unhurt (*ibid.*: 41–91).

The crew knew generally but not exactly where they were. The radio had been smashed and as the winds howled over them Monteverde gathered the crew together and announced: "According to Regulations, I am in charge, but I want any and all suggestions you might happen to think of. We will work it out together" (Balchen *op cit.*: 47; see also: Taub, n.d.). The men huddled into the wrecked and cramped fuselage, and it was not until November 24 that they heard a large C-54 transport fly overhead, with Balchen on board. He had been searching for the aircraft ever since it became clear the B-17 was down. The C-54 flew just 200 feet overhead, dropping ration packs, a stove, sleeping bags, and proper Arctic clothing.

Once the plane had been located, Max Demorest set off with his two new motor sledges, each towing a sled full of supplies. The other sledge was driven by a Staff Sergeant, Don Tetley of San Antonio, Texas, who had only arrived in Greenland and seen Arctic ice for the first time 3 months earlier. Together they became the first to travel across the Greenland Ice Cap in winter.

At the same time, Lieutenant John Pritchard, Jr., of the Coast Guard cutter *Northland*, took off in a Grumman Duck floatplane with his lone crewmember, Radioman Benjamin Bottoms. Pritchard made a wheels-down landing near the crash site and successfully evacuated two of the B-17 crew back to the *Northland*.

Soon after, Demorest and Tetley arrived, skiing ahead of their sledges. Pritchard was due back the next day. The ordeal would soon be over.

In the morning, however, Demorest and Tetley returned to the sledges, to bring them up to the downed B-17, when Demorest fell through the ice and disappeared into a crevasse. Tetley searched in vain and the B-17 crew brought up rope to make a rope ladder, but the crevasse was impossibly deep, and no sign of Demorest was ever seen or heard from again.

Pritchard and Bottoms returned that morning, taking away another crewmember from the B-17, but their Grumman itself disappeared in a growing storm and all three on board were lost. Two men had now been rescued at the cost of four. The downed aircrew continued to receive air drops of supplies. A dog sled team sent from the eastern coast was also stuck on the Ice Cap and relief supplies now had to be dropped to them, along with Tetley at his sledge and the remaining men of the B-17.

It was not until early February of 1943 that conditions allowed Balchen to circle the crash site in his Catalina, along with a second Catalina and a B-17 for support. Balchen made a wheels-up belly landing on the ice. As the downed airmen were loaded onto the flying boat, the hull became stuck to the ice. Balchen directed the men to waggle the wingtip floats until the aircraft could begin to move, at which time it maneuvered in circles to keep from freezing again, as one by one the downed airmen were hauled on board through the open gun blisters on the side of the fuselage. At the end, five servicemen were dead and one had to have both feet amputated. Four had survived uninjured.

During their three-month ordeal on the Greenland Ice Cap, the tide of the global war in which the downed B-17 crew had been engaged had turned. In November, during their first days on the ice, British general Bernard Montgomery's Eighth Army had defeated the Germans under Rommel at the Second Battle of El Alamein. On the day they were rescued, the German 6th Army had been destroyed at Stalingrad. By the spring of 1943, the battle against German U-boats in the Atlantic would turn as well, and with it the need for Greenland as the primary route for fighter aircraft to be flown to England. The planes could now also be carried in greater numbers on board transport ships, along with the massive amounts of supplies required for the invasion of Nazi-occupied Europe in June of 1944. Besides fighter aircraft, the largest four-engined bombers, with favorable winds, could make the hop from Newfoundland to Scotland in one go and avoid Greenland altogether.

As the war turned, Balchen moved on as well, to even more secret missions for General Arnold. These included the resupply of the Norwegian Resistance by his squadron of B-24 Liberator bombers. His Norwegian-accented English whenever offered an extreme mission became the unit's motto: *'Ve Do It.'*

The focus elsewhere allowed Iceberg Smith to keep one or two of the units under his Greenland Patrol command at work engaged in hydrographic surveys of Greenland's coastal and fjord waters throughout the war. Aids to navigation, a loran station, radio beacons, light stations and day marks, along with the major port facilities at Bluie West 1, all became after the war the maritime transport infrastructure of Greenland. Bluie West 1 itself had grown to include a camp for 70 officers and 310 enlisted men, along with Smith's administrative headquarters for the Greenland Patrol, a maintenance base for PBY Catalinas, a radio station, and a mooring large enough to handle a U.S. Navy cruiser (U.S. Navy, 1947: 60).

By 1943, the three weather stations hurriedly established along the Greenland coast in 1941 by an aviation meteorology cadet from New York University named Robert B. Sykes had grown to twenty-six, and by June of 1944 helped provide General Dwight D. Eisenhower with the weather forecasting required to set the date for D-Day (Fuller, 1990: 55; 61).

Iceberg Smith had been promoted to Rear Admiral in 1942, a rank that would be made permanent after the war. But then Smith's career stalled. He found himself a specialist in a service of all-arounders. The expertise in the ice that had been required to defend Greenland during the race to fortify the island between 1940 and 1942 was now a luxury to a service that shrank to a fraction of its war-time size. Smith was assigned to be the Captain of the Port of New York, a billet that allowed enough spare time to serve on the board of the Woods Hole Oceanographic Institution, where he became director following his retirement from military service in 1950.

For Balchen, no such crowning position awaited. The U.S. Army Air Corps had become the U.S. Army Air Force and then, in 1947, a separate military branch entirely, the U.S. Air Force. His mentor and champion Hap Arnold, who guided the separation of the Air Force from the Army and thus became the only American officer to hold five-star rank in two services, died in 1950. Balchen was put forward repeatedly for promotion to the rank of general but was blocked, so Balchen believed, by the jealousies and insecurities of Richard Byrd, and Byrd's well-connected family and friends in Washington, D.C. Balchen's post-war fame, derived in good measure from the dramatic ice rescues atop Greenland, had particularly rankled the publicity-obsessed Byrd. As a naval officer, Byrd had long fought to suppress both an independent air arm and most especially its use in any Arctic work. As a world-famous Arctic expert—and a senior officer in the newly-formed U.S. Air Force into the bargain—Balchen represented everything Byrd despised. After a dramatic turn as commander of a search and rescue base in Alaska, Balchen was reduced to recruiting tours and radio and television appearances when he wasn't marking time in a shared basement office in the Pentagon.

For a brief moment, it seemed that Greenland itself was set to return to its relative northern obscurity. Then, just as the war reached its climax, as Soviet armies raced across northern Germany and towards Denmark, Allied forces under Montgomery in early May reached the Baltic port of Wismar and the city of Lübeck just hours ahead of the Russians. Denmark was therefore saved from Soviet occupation, and all the attendant effects such an occupation would have meant for Greenland.

One war had ended. Another was about to begin. And Greenland once again would be in the middle of it.

References

Balchen, B.. (with Corey Ford) (1945). *War below zero: The Battle for Greenland.* George Allen & Unwin.

Capelotti, P. J. (1996). *Oceanography in the coast guard.* U.S. Coast Guard Historians Office.

Capelotti, P. J. (Ed.). (2005). *Life and death on the Greenland patrol.* University Press of Florida.

Carlson, W. S. (1962). *Lifelines through the Arctic.* Duell, Sloan and Pearce.

Desh, R. (2023). *The North-East Greenland sledge patrol: A unique and lasting legacy of the U.S. Coast Guard's defense of Greenland during WWII.* U.S. Coast Guard Historians Office. Online at: Retrieved February 4, 2026, from https://media.defense.gov/2023/Sep/26/2003308608/-1/-1/0/DESH_NE_GREENLAND_SLEDGE_PATROL.PDF

Fuller, J. F. (1990). *Thor's legions: Weather support to the U.S. Air Force and Army, 1937–1987.* American Meteorological Society.

Glines, C. V. (1999). *Bernt Balchen, Polar Aviator.* Smithsonian Institution Press.

Hansen, C. (2012). *Enfant terrible: The times and schemes of General Elliott Roosevelt.* Able Baker Press.

Hobbs, W. H. (1952). *An explorer-scientist's pilgrimage.* J.W. Edwards.

Morison, S. E. (1947). *The Battle of the Atlantic.* Little, Brown.

Smith, E. H. (1949). *The U.S. Coast Guard in the Arctic and Sub-Arctic. 13-p. typescript, written for Encyclopedia Arctica.* U.S. Coast Guard Historians Office.

Taub, D. (n.d.). *The Greenland ice cap rescue of B-17 "PN9E".* U.S. Coast Guard Historians Office. See also: Guldager, Ole. 2019. Americans in Greenland in World War Two. Arctic Sun: Greenland History Series.

U.S. Navy. (1947). *Building the navy's bases in world war II, volume II, part III: The advance bases.* Government Printing Office. Online at: Retrieved February 3, 2026, from https://www.history.navy.mil/research/library/online-reading-room/title-list-alphabetically/b/building-the-navys-bases/building-the-navys-bases-vol-2.html

Wilford, J. N. (1996). Did Byrd reach the pole? His diary hints 'no. *New York Times, 9*, 1996.

Chapter 10
Bernt Balchen and the Post-War Creation of Greenland as a NATO Island

Abstract In the aftermath of the Second World War, the U.S. attempted once again to acquire Greenland and, that failing, exert control over military operations on the island. This led to increasingly daring and dangerous proposals, ending with the failed experiment at Camp Century and the cancelled Project Iceworm. As the Greenlandic people began to demand more control over their own destiny, and both an historical reckoning and an environmental catastrophe loomed, tensions between the U.S., Denmark, and Greenland intensified.

Keywords Greenland · Greenlandic Inuit · Thule · 1949 North Atlantic Treaty · U.S. Army · U.S. Air Force · 1951 Agreement · Distant Early Warning system · Pittufik Space Base · U.S. Coast Guard · U.S. Navy · Camp Century · Project Iceworm · Thule Affair · Greenlandic Home Rule · Cryolite

> "During the warm months at Qangattarsa, polar bears sometimes came to the dump to forage. The bears were a nuisance and dangerous, so the dump was off limits, but occasionally the colonel authorized Periodic Ursine Harassment Details (PUHDs)…
>
> "Several Jeeps and a half-track were chasing the two white bears around smoldering heaps of garbage…
>
> 'There's something very Rome, four AD, about all this,' Lavone said. 'An amphitheater. Beasts. Soldiers in chariots.' He raised his beer aloft. 'Emperor Caius Augustus Woolwrap, we salute thee.'
>
> —John Griesemer, *No One Thinks of Greenland* (2001)

Winston Churchill delivered his Iron Curtain speech on March 5, 1946, at Westminster College in Fulton, Missouri. But the Cold War had begun a year earlier when Bernard Montgomery's forces raced to close the Danish peninsula to the advancing armies of Soviet Russia. Post-war Russian control of Denmark would have thrown Greenland once again onto the world stage. In the event, it happened anyway.

P. J. Capelotti, *Lost Greenland*, Springer Polar Sciences,
https://doi.org/10.1007/978-3-032-27841-8_10

Montgomery himself was soon appointed Chief of the Imperial General Staff and, in 1948, Commander in Chief of the Western Union Defence Organization. This forerunner of the 1949 North Atlantic Treaty Organization contained, like NATO, guarantees of mutual aid in the event of an attack on any of the five nations involved: the United Kingdom, France, Belgium, Luxembourg, and the Netherlands. Montgomery then served for a decade as Deputy Supreme Allied Commander Europe, in his first year as deputy to the first NATO commander, Dwight Eisenhower.

The Allied salvation of Denmark from decades behind the Iron Curtain contained within it the seeds of continued American interests in Greenland. With the threat of Germany replaced by the new threat from an expansionist Soviet Union, Greenland once again became the key to the defense of the Western Hemisphere. For some in the leadership of the U.S. military, actual ownership of that key became their goal. Just one month after Churchill's speech, a planning meeting of the American Joint Chiefs of Staff concluded that "our real objective as regards to Greenland should be to acquire it by purchase from Denmark," a notion that apparently originated with a U.S. Senator from Maine and went unrevealed for nearly half a century until it was reported by the Copenhagen newspaper *Jyllands-Posten* in the spring of 1991 (see: Associated Press, 1991).

A purchase price was even meted about in May of 1946 by an assistant chief in the U.S. State Department: $100 million in gold (*ibid.*), an offer seconded by the Secretary of War Robert Patterson in June. The proposal was made formally by Secretary of State James Byrnes to the Danish Foreign Minister Gustav Rasmussen in December of 1946, who received it with surprise and shock.

There is no evidence the Danes ever responded formally to the offer, but it did signal that the U.S. in the aftermath of the Second World War was not leaving Greenland easily, if at all. All U.S. policy in the immediate post-war period, from the 1947 Truman Doctrine to provide support to democratic nations under threat from communism to the 1948 Marshall Plan to rebuild Europe, to the formation of NATO in 1949, began with the premise that the United States was both sword and shield. No other nation could project force around the globe as could the U.S. at the same time as it invested billions to make certain a Pearl Harbor-like attack on America could never happen again. Such a defensive shield relied on advance bases in Greenland, early warning radar that could provide crucial extra minutes of alert that could spell the difference between survival and atomic annihilation.

As the Cold War entrenched itself on both sides of the Atlantic, the notion of Greenland as an early warning tripwire was expressed in both official policy, military preparations, and popular culture that veered into the realm of science fiction. A 1950 article in *Mechanix Illustrated* proposed "a string of giant rubber bubbles, housing radar sentries, hidden in the icy peaks of America's northernmost mountains" while making clear that 'America's northernmost mountains' included those in Greenland (Tinsley, 1950: 82–83) (Fig. 10.1).

As the general post-war military drawdown proceeded, such proposals seemed to gain a permanent foothold in Greenland. The many American military bases on the

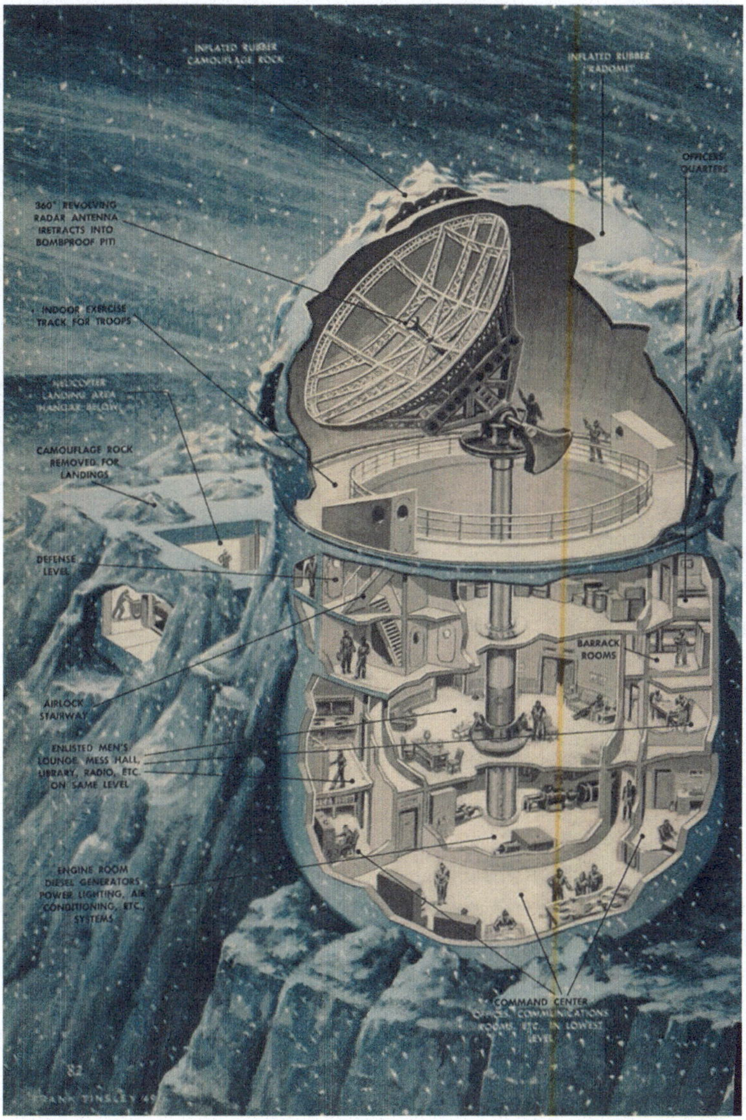

Fig. 10.1 An imagined inflatable radome proposed to be sited within a Greenland mountain, from a 1949 article in *Mechanix Illustrated* (from the author's collection)

island underwent either consolidation, elimination, or redirection away from aviation support to early warning radar facilities (Fig. 10.2).

Bluie East 2, as one example, was quickly closed down in 1947 after just 4 years in operation, its dozens of vehicles abandoned and left to rust in place, while the tens of thousands of fuel drums at the site were allowed by the U.S. to contaminate the waters of the adjacent Ikateq Fjord for decades, destroying the ecosystem that

Fig. 10.2 The collapsed hangar at Bluie East 2, in July 2025. The dish antenna mounted on a trailer is possibly a very early example of weather radar trailer. If true, it would mark one of the earliest surviving examples anywhere (photo by the author)

had sustained "customary hunting and fishing grounds and a campsite of the Inuit of the nearby villages" (Elixhauser, 2019: 115). Under the terms of the 1951 Defense of Greenland Agreement between Denmark and the U.S., these hazardous materials should have been removed "within a reasonable time" (Defense of Greenland, Article XI) but were still *in situ* into the twenty-first century.

As the wartime control of Greenland's bases passed in the immediate postwar period from the U.S. Army to the newly chartered U.S. Air Force, Denmark found itself in the awkward position of having to accommodate a guest that showed no inclination to leave. The U.S. had built military support structures in Arctic Canada during the war, and these reverted largely to Canadian control in the aftermath. Denmark, like the rest of Europe prostrate from six years of war, was in no position to challenge the post-war dictates of the U.S., the sole atomic nation (until 1949) to emerge from the war and the only one largely unscathed and with its financial power and prestige and global military reach intact. If the U.S. wanted to continue operations in Greenland, even in a more focused manner, Denmark would have to adjust to this reality.

The 1951 agreement between Denmark and the U.S. over military bases in Greenland, signed in Copenhagen on April 27, was predicated on the earlier North Atlantic Treaty both countries had signed two years earlier in Washington. Under the terms of the NATO Treaty, the two countries now agreed that either nation "may make use of facilities in Greenland in defense of Greenland and the rest of the North Atlantic Treaty area" (Defense of Greenland, Preamble). In return, the U.S. in

August of 1951 turned over to Denmark control of Bluie West 7, the naval base at Grønnedal (Kangilinnguit) that had grown out of Iceberg Smith's plan for the defense of the Ivigtut cryolite mine. Denmark also took control of the weather reporting stations that had been operated by the United States.

As in the nineteenth century, little or no thought was given to the indigenous Greenlanders themselves. As Charles Emmerson has written, in words as easily applied to the twentieth century: "The idea of native sovereignty over Greenland did not seriously occur to the lawyers in the nineteenth century. For them, the transfer of Greenland from one European people to another was entirely valid" (Emmerson, 2011: 103). Postwar changes were coming, but for many they were too slow, too incremental, and too focused in the colonizers, not the colonized. Associated with the Norse for a millennium and a Danish colony for 250 years since Hans Egede, Greenland "was redefined as a district of Denmark" in 1953 (see: Kingdom of Denmark n.d.). Its local government had two representatives in the *Folketing*, the Danish parliament.

The major impact of the 1951 U.S.-Denmark agreement (Anon., n.d.) came much further north. Besides leading an Arctic search and rescue command in Alaska after the war, Bernt Balchen's other main contribution came in this same year of 1950, when he recommended Thule in northwest Greenland as the site for a strategic air base and listening station. No less than Knud Rasmussen had advised Balchen in 1927 that Thule was far superior to Svalbard for a launch point for aircraft towards the North Pole (Glines *op cit.*: 143). Balchen remembered this when he was in command of Bluie West 8 during the war, and in August of 1942 made a survey flight north to check on the area. He came away impressed and advocated for a US base at Thule as early as 1942.

Now, in the rush to counter the Soviet Union, Balchen went much further. From his basement office in the Pentagon, Balchen "submitted a paper to Air Force Secretary Stuart Symington proposing a central search and rescue organization for all northern military operations calling it a "sort of Arctic NATO." He envisioned a base constructed at Thule as a centerpiece for these Arctic operations along with serving as a base for military aircraft and civilian aircraft emergencies" (Swain, 2024: 14). Ultimately, the Air Force Chief of Staff assigned Balchen as project officer for Operation Blue Jay, the secret construction of what became Thule Air Base and since 2023, in recognition of its Greenlandic cultural heritage, Pittufik Space Base. The same area where Cook and Peary had warred over Annoatok Island, where Charles Francis Hall was buried and Elisha Kent Kane had lost his ship, would now become an early warning radar location on guard against Soviet Russian missiles and bombers.

Balchen arrived on site in February of 1951 where his surveying party faced temperatures of -40F and 100 knot winds. Ten thousand men, all recruited from cold weather states, along with 300,000 tons of equipment, joined them in mid-summer and by September of 1951 a 10,000-foot runway was ready to accommodate long-range strategic bombers. "Beside the massive runway, the base included barracks, fuel storage, offices, warehouses, heated hangars, the world's largest saltwater distillery and the second tallest radio tower. With this level of construction, Thule was not long a secret" (*ibid.*: 15). The secrecy ended officially a year later, and Balchen accompanied three dozen journalists from Washington, D.C. on the visit to see the new base.

Behind the scenes, Balchen was secretly evaluating NATO requirements should war break out with the Soviet Union. His fame and language skills allowed him access to intelligence reports from throughout Fennoscandia, the region of the Scandinavian peninsula as well as Finland and northwestern Russia. Balchen wrote up an operational evaluation of the defense of the peninsula and Denmark in the event of a Soviet attack, and soon became heavily involved in the creation of the Distant Early Warning (DEW) Line of radar stations stretching from Alaska to Greenland were charged with detecting Soviet bombers before they could cross the North Pole (Glines *op cit.*: 229–230).

In 1958, the DEW Line sites were also extended along the East Greenland coast and into Iceland (the DYE sites), though Balchen had advocated for the line to be set much further north, along Greenland's northern polar coast. It was Balchen who secured agreement with Norway to allow the U.S. to establish early warning sites on Norwegian soil.

It was soon after this mission that Balchen was sidelined at the Pentagon from more Arctic work. He never did learn why he'd been pulled from work in the north but assumed it was the work of Richard Byrd. Whatever the cause, it did not prevent President Dwight D. Eisenhower from awarding Balchen the Harmon International Trophy in November of 1953, "for outstanding service to the advancement of aviation in connection with Arctic operations, Arctic exploration, Arctic rescue and Arctic pilotage, including many polar flights" (*ibid.*: 240). With the creation of Thule, Balchen had secured Greenland as a NATO island and, along with Norway itself, the front line of defense against Soviet communism.

The Coast Guard, for its part, continued its operations off the Greenland coast. In the summer of 1950, the icebreakers *Eastwind* and *Edisto*, accompanied by the cargo vessel USS *Whitley* and the LST 533, resupplied the several joint weather stations including the one at Alert on Ellesmere Island. During the operation, *Eastwind* traveled north through thirty feet of pack ice to reach with 445 miles of the North Pole, the northernmost point reached to that date by any surface vessel under its own power. The mission transported "Canadian government surveyors and other field technicians to locations where astro fixes, terrain elevation, geology, wildlife and other studies were accomplished" (Peterson, 1951: 8–9).

Coast Guard operations turned increasingly to the resupply of Thule, where excellent harbor facilities now complimented the heavy airlift runway. In 1956, Walt Disney released a film, *Men Against the Arctic*, that employed Chief Photographer Elmo Jones to document a season's cruise by the Coast Guard's heavy icebreakers *Eastwind* and *Westwind* as they resupplied the new Thule base as well as the weather station at Alert. "Coast Guardsmen .... have conquered the Arctic and have made it their own. Not once, but many times, they have turned northward toward the ice and desolation and the infinite loneliness to prove again their daring and their fortitude" (Pollard, 1956). In 1964 the Coast Guard assumed all icebreaking duties for the United States and the navy transferred all its remaining ice-capable ships to the Coast Guard. For a moment, there seemed nothing the U.S. military could not do in the North, and soon they set out to prove it.

The renewed focus on the Arctic came about by several sudden sharp shocks to the relative international equilibrium since the announcement of the Soviet acquisition of atomic weapons in 1949. In October of 1957, the somewhat predictable world of atomic bomb-equipped long-range aircraft and their associated bases and listening stations developed by Bernt Balchen and others in the Arctic was overturned by the launch of the first artificial Earth satellite, Sputnik 1. Overnight the threat of atomic destruction mutated from bombers to missiles, for which no defense existed and against which no early alert system could provide more than a few minutes of warning.

The following year, the seeming Soviet technological advantage was challenged by the U.S. Navy and its voyage under the ice to the North Pole of the nuclear submarine USS *Nautilus*. This was followed one year later by the submarine USS *Skate*, which surfaced at the pole. Silent, secret, missile-equipped submarines could now lurk anywhere in the world's oceans, ready to strike back against any missile attack by the Soviets. The infamous acronym MAD (Mutually Assured Destruction) became the new military doctrine that guaranteed that any first strike against another atomic power would be met by the near simultaneous annihilation of the attacker. Hiding a nation's missile counter-strike options was now imperative.

Accordingly, the most ambitious American projects in Greenland came in 1959, with the construction of Camp Century and its highly classified offspring, Project Iceworm. Nominally a 'secret base under the ice,' the U.S. Army Corps of Engineers Camp Century project was soon featured on the cover of *Popular Science* in February of 1960, and CBS newsman Walter Cronkite was flown to the base to film a special in the CBS "The Twentieth Century" series. The writer of the documentary film, Walter Wager, converted his script into a popular account published in 1962: *Camp Century: City Under the Ice* (Wager, 1962).

To create the under-ice facility, nearly 10,000 feet of tunnels were excavated into the Greenland Ice Cap about 200 km (100 miles: hence the name 'century') east-northeast of Thule Air Base. Into these tunnels were placed prefabricated structures including dining, bathroom, and medical facilities. The whole area was powered by a steam turbine driven by a small uranium nuclear reactor. It was, as Wager wrote: "an exceptional human achievement … a dramatic tribute to the imagination, courage, stubbornness and skill of the U.S. Army Corps of Engineers… the entire project was made possible by the cooperation of our North Atlantic Treaty Organization partner, Denmark. Greenland is a province of the Kingdom of Denmark, and American forces are on the chill northern island under the terms of a treaty between Denmark and the United States" (*ibid.*: 2–3).

Ignoring a century of American explorations in and around Greenland and not least the inter-war work of Hobbs, Cramer, Boyd, Lindbergh, and MacMillan ("There was little U.S. interest in Greenland between 1916 and 1941" (*ibid.*: 12)), Wager framed Camp Century in the language of Cold War science, "a home for men seeking further knowledge about the Arctic. Dedicated American civilian scientists are trying to discover how strong snow is, how to build with it, how to live on it, how to travel over it, how to make airplane landing strips of it, how to construct huge unlined warehouses and storage tanks in it, and how to study it for the history of our planet's weather" (*ibid.*: 5). Wager failed to mention that the U.S. had not asked permission from the Danish government to install a nuclear reactor under the Greenland ice (Nielsen & Nielsen, 2017: 34–35) (Figs. 10.3 and 10.4).

In an echo of the earlier Greenland propaganda expeditions, it also became the home for six months of two boy scouts, one from Denmark and one from the U.S., selected in a competition helmed by Paul Siple, who had been similarly chosen as a scout to accompany Richard Byrd's expedition to Antarctica in 1928 (*ibid.*: 144–174). The Danish scout even learned secrets that would go unacknowledged by the U.S. for decades: "Concerning the disposal of radioactive waste, ... permission has been given by the Danish government to pour up to 50 millicuries of liquid waste out into the ice per year" (*ibid.*: 157).

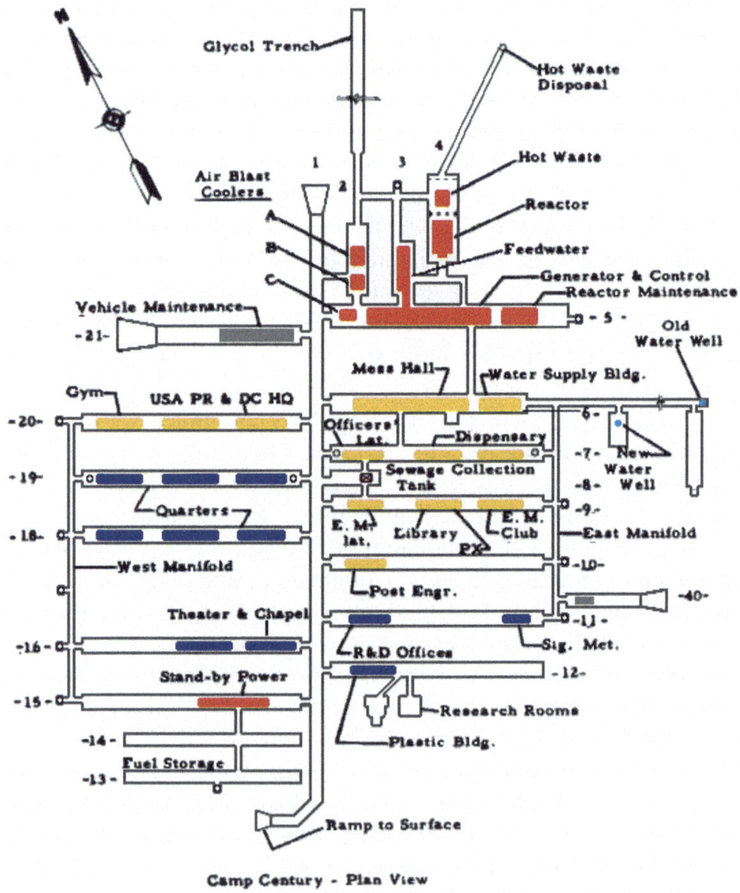

Fig. 10.3 Plan view of Camp Century (public domain via the U.S. Army Corps of Engineers)

Figure 14. Metal arch forms being emplaced over undercut trench.

Fig. 10.4 Metal arches forming the roof over once of the Camp Century trenches (public domain via the U.S. Army Corps of Engineers)

Unbeknownst to Wager or Cronkite or anyone else, including the two Boy Scouts, Camp Century, elaborate though it was, was only a cover for a much larger, and infinitely more classified, scheme: Project Iceworm, a proposal to build a network of nuclear missile launch sites under the Greenland ice that could survive a first strike against North America and NATO. Where Camp Century was about one mile of tunnels, Iceworm envisioned a system of more than *2500* miles (4000 km) of Ice Cap tunnels where up to 600 Medium Range Ballistic Missiles (MRBM) would be moved in response to imminent threats, similar to the 1970s proposal for a railroad track-based MX missile transport system.

As with Century, the true purpose of Iceworm was hidden, in particular from the Danish government itself, for decades, the original documentation not declassified until 1996 (Petersen, 2007). As the age of strategic bombers began to give way to a greater reliance on intercontinental missiles for nuclear deterrence, Iceworm was one of three entries into the transcontinental missile race. This inter-service rivalry pitted the army's Iceworm project against the U.S. Air Force's Minuteman and the U.S. Navy's Polaris. The navy and air force ultimately prevailed and, combined with the desultory results regarding Camp Century's unstable structural integrity after just a year or two, led to the army's cancellation of Iceworm and the Air Force assuming the primary U.S. military presence at Thule Air Base.

Camp Century was scheduled to close soon after, and with it the removal of its nuclear reactor. The Army cold weather research station at Camp Fistclench, used for tests prior to Century, ended in 1960, and Camp Tuto, a similar cold weather research outpost at the base of the Ice Cap, ended in 1966. The army revisited Century once again, in 1969, only to find that the tunnels had nearly completely collapsed. "Today, there are no visible traces of the camp on the surface" (*ibid.*: 244).

The legacy of Camp Century was primarily two-fold. The first came in 1966, when an ice core drill at the camp touched the bedrock at the bottom of the ice sheet. This ice core allowed for the first time an analysis of the planet's atmosphere going back thousands of years, and "marked a turning point in geophysical research in Greenland and a shift in our understanding of Greenland" (*ibid.*: 222). From this point, research in Greenland would shift from the army and mostly military operations to the National Science Foundation's collaborative projects with other circumpolar nations, on research primarily directed at questions of global climate change.

The second legacy of Camp Century was one that will linger on perhaps unto the twenty-second century: the slow inexorable movement of the radioactive camp from its location now one hundred feet or more beneath the ice sheet out to the waters off northwestern Greenland. The final army visit to Century in 1969 followed by just one year the crash on the frozen surface of Wolstenholme Fjord near Thule Air Base of a hydrogen bomb-carrying U.S. Air Force B-52 Stratofortress. Six of the seven crew members ejected safely from the aircraft, but the impact set off the B-52's conventional munitions to detonate, rupturing the four hydrogen bombs and scattering radioactive contamination in the area. In what became known as the Thule Affair (*Thuleulykken*): "Several hundred people from the United States, Denmark, and Greenland spent months collecting snow and ice contaminated with plutonium and depositing it in large containers that were subsequently transported to the United States" (*ibid.*: 252). A similar incident occurred two years earlier off Spain. Then, three bombs had dropped on land but the fourth fell into the Mediterranean and resulted in a dramatic three-month search by multiple submersibles before the bomb was recovered.

The rapid mobilization of local Greenlanders once again coming to the aid of an American expedition held both historical echoes and future worries. Decades later, reports of increased cancers amongst the Greenlanders who participated in the 1968 clean-up led to distrust of official Danish explanations of the relative safety of radioactivity levels near Thule. News reports, books, and documentaries followed that only intensified feelings in both Denmark and Greenland that authorities had for decades concealed information on U.S. nuclear-armed flights across Greenland and the assumed storage of atomic weapons at Thule.

This growing distrust took place against the backdrop of increasing calls for greater autonomy for Greenland from Denmark, if not outright independence. Home Rule was introduced in Greenland in 1979, following a quarter century of de-colonization of Greenland by Denmark through the 1953 change to the Danish Constitution that also resulted in a long attempt at the 'danification' of Greenland

(Jakobsen & Larsen, 2024: 13). The Thule Affair only added to local anger at the differential treatment of Greenlanders versus Danes in Greenland in such basic areas as salaries and control of Greenlandic natural resources. A 1972 vote to join the European Community was approved by a vote of 63%–37% in Denmark, but in the 'Danish county' of Greenland the vote was the opposite, 71% opposed to only 29% in favor (*ibid.*: 14).

In response, a Home Rule committee formed in Greenland, and the Greenland Home Rule Act followed into force on May 1, 1979. A 1995 Danish parliamentary report on the Thule Affair led as well to a reexamination of Camp Century, which in turn led to the disclosure of the clandestine Project Iceworm. A media firestorm ensued and, along with it, a renewed assessment of the complex relationship of Greenland within the wider relations between Denmark and the United States, and between the U.S. and all NATO allies. Soon after, Kenn Harper's brilliant *Give Me My Father's Body* returned the story of the boy Minik from Northwest Greenland to the same front pages where it had been relayed to the American public a century before. But this time the tale was told in a very different way.

> "In (Robert Peary's) writings, the Eskimos were "my faithful, trusty Eskimo allies, dusky children of the Pole" and "effective instruments for Arctic work." At the end of his Arctic career, after an association of almost twenty years, he had the arrogance to write that "these people are much like children, and should be treated as such."...
>
> "To be sure, some Eskimos genuinely liked the man. The influential Uutaaq was among them. But many others retain a quite different memory of him. In 1967 an elderly man in tiny Siorapaluk, the district's most northerly village, reminisced about Peary, the man he referred to as "the great tormentor":
>
> "People were afraid of him ... really afraid.... His big ship ... it made a big impression on us. He was a great leader. You always had the feeling that if you didn't do what he wanted, he would condemn you to death.... I was very young, but I will never forget how he treated the Inuit.... His big ship arrives in the bay. He is hardly visible from the shore, but he shouts: 'Kiiha Tikeqihunga! — I'm arriving, for a fact!' (Harper, 2000).

As Greenland entered the twenty-first century, it remained burdened by its long and difficult relationship not only with its one-time colonial administrator, Denmark, but by the bewildering variety of American explorers, entrepreneurs, and military planners who could all with some justice be gathered into one 'great tormenter.' From Elisha Kent Kane and Isaac Israel Hayes helping themselves to Greenlandic grave goods in the nineteenth century, to Peary's and Henson's use of Greenlandic men for guides and women for concubines in the 20th, to the lingering dangers of atomic bases and radioactive contaminants of the 1960s, Greenlanders themselves had precious little say in how America and Americans used and abused their island.

Greenlandic Home Rule became an Act of Self-Government in 2009, and with it Greenlandic control of mineral resources and the offshore environment, albeit with foreign and defense policies still decided in Copenhagen. Moves towards full independence stalled in the absence of full Greenlandic control over a national bureaucracy and the national income to keep it functioning, combined with the threat of what would surely be the loss of all Danish subsidies. In 2014, Greenland established a reconciliation committee, but Denmark refused to participate (Jakobsen and Larsen *op cit.*: 23.).

When a 2016 study published in *Geophysical Research Letters* found that Camp Century with its tonnage of raw sewage and radioactive waste could emerge from the Greenland Ice Cap as early as the year 2090, Denmark and Greenland responded with the formation of 'The Camp Century Climate Monitoring Programme.' A permanent research base located just above the ruins of Camp Century was constructed in the summer of 2017, specifically to investigate the environmental, health, and pollution challenges that will surround the buried base until such time as its structures and associated hazardous wastes emerge from the ice (Nielsen and Nielsen *op cit.*: 273).

Reminders of the American legacy were not limited to the west coast of Greenland. The Danish government announced in 2017 that it would fund efforts to clean up pollution at U.S. military bases abandoned in Greenland more than half a century before, including Bluie East 2. The six-year, 180 million kroner ($30 million) program involved the establishment of a group of environmental experts from Greenland and Denmark who would work with local residents to identify the extent of pollution at the World War II-era installations and then prioritize which sites should be cleaned first. The agreement came only after Nuuk had long pressed Copenhagen to either clean up the pollution itself, or to request that Washington, which Denmark permitted to establish the bases in Greenland during the Second World War, do so (McGwin, 2018) (Fig. 10.5).

American interest in Greenland made an almost comic reappearance in 2019 when businessman, former U.S. ambassador to Austria, and self-described 'Greenland expert' Ronald Lauder suggested to President Donald Trump that the U.S. purchase Greenland from Denmark. This was followed by the return of a U.S. consulate to Nuuk in the summer of 2020, after an absence of nearly 70 years. The first U.S. consulate, opened in Godthåb (Nuuk) after the Nazi invasion of Denmark in 1940, closed in 1953. These moves, some serious, most not, were at best half-hearted responses to China's aggressive 'Polar Silk Road' initiative, combined with Russian moves to reinforce its own Arctic frontier through the expansion of bases in Franz Josef Land and repeated probes into the seriousness of Norwegian sovereignty over Svalbard.

With the 2020 U.S. presidential election, Greenland once again faded into the background, as the practical difficulties of extracting rare earth elements there chastened corporate interest in expanding operations beyond the existing concerns like the historic cryolite mine at Ivigtut. But with the 2024 election, Greenland emerged once again, and loudly. The question now became whether the consulate in Nuuk remains just that or perhaps becomes the U.S. embassy to an independent Greenland or—as Trump himself repeatedly suggested—is recast, by force, if necessary, as the capital of an American colony in the Arctic. As one commentator put it: "The apparent sense of freedom that Trump and (Elon) Musk feel in playing politics abroad is also a marker of the self-confidence … ahead of Trump's inauguration. They're demonstrating a belief that their strength allows them to bully smaller countries and may augur a new and brasher incarnation of "America First."" (Collison, 2025).

Fig. 10.5 USCGC *Campbell* in 2020 transiting south along the west coast of Greenland with the Royal Danish navy vessel HDMS *Knud Rasmussen*. The two vessels rendezvoused in a position just offshore of Evighedsfjorden (Eternity Fjord), where *Campbell* received *Rasmussen's* Executive Officer, Commander Bo Ougaard, on board to serve as ice pilot and provide local knowledge to assist *Campbell* in safely entering and transiting the fjord. Once inside Eternity Fjord, CGC Campbell launched their MH-65 Dolphin aircraft and proceeded up the fjord to the head where the glacier begins (public Domain: U.S. Coast Guard photo by Seaman Kate Kilroy)

The reasons behind the apparently bizarre fixation were not difficult to find. It lay in the belief that beneath the Greenland Ice Cap lay riches beyond the dreams of avarice, and for a new class of American oligarchs such dreams had reached fever level. "Like many of those around the president, (Ronald) Lauder's policy suggestions appear to intersect with his business interests. As Trump has ratcheted up his threats to seize Greenland, Lauder has acquired commercial holdings there. Lauder is also part of the consortium whose desire to access Ukrainian minerals appears to have spurred Trump to demand a share of the war-torn country's resources" (Burgis, 2026).

Such neo-colonialist ideas struck many as throw-back to the nineteenth century and William Seward's scramble for Alaska, the Virgin Islands, and Iceland, as well as Greenland, along with America's later Spanish-America War spoils of Puerto Rico, Guam, the Philippines, Cuba, and the annexation of Hawai'i. American comedian Conan O'Brien even flew to Nuuk to 'negotiate' the sale of Greenland with Pele Broberg of the Greenland Parliament. Broberg thanked O'Brien for at least coming to Nuuk and not Copenhagen to open his 'negotiations.'

O'Brien then offered bribes consisting of a Starbucks card and a suitcase full of $120,000 in small bills. O'Brien then learned firsthand how a Greenlander would respond to such inducements when Broberg looked dismissively at the cash and asked: "And you did of course pay taxes importing it to Greenland?" (Team Coco, 2019). If nothing else, the spoof, offered amid an environment of threatened U.S. military coercion and implied commercial corruption, demonstrated that the American understanding of Greenland and Greenlanders had not come far since Ben Franklin's 1739 *Poor Richard's Almanac* rhapsody on 'the Orange Trees in Greenland.'

References

Anon (n.d.). *Defense of Greenland: Agreement Between the United States and the Kingdom of Denmark, April 27, 1951*. Avalon Project, Yale University, online at: Retrieved February 5, 2026 https://avalon.law.yale.edu/20th_century/den001.asp#:~:text=In%20order%20that%20the%20Government%20of%20the,of%20America%2C%20the%20following%20provisions%20shall%20apply

Associated Press. (1991). *The United States in 1946 proposed to pay Denmark $100 million to buy Greenland*. Online at: Retrieved February 4, 2026, from https://apnews.com/article/9d4a8021c3650800fdf6dd5903f68972

Burgis, T. (2026). How a billionaire with interests in Greenland encouraged Trump to acquire the territory. *The Guardian*. January 15, 2026. Online at: Retrieved February12, 2026, from https://www.theguardian.com/us-news/2026/jan/15/ronald-lauder-billionaire-donor-donald-trump-ukraine-greenland

Collison, S.. (2025). *Musk plays politics abroad as world leaders brace for Trump's return*. CNN.com, January 7, 2025. Online at: Retrieved February 12, 2026, from https://www.cnn.com/2025/01/07/politics/the-dangers-of-elon-musks-personal-foreign-policy/index.html

Elixhauser, S. (2019). Inuit responses to Arctic militarization. In Herzberg, Kehrt, & Torma (Eds.), *Ice and snow in the cold war*. Berghahn.

Emmerson, C. (2011). *The future history of the Arctic*. Vintage Books.

Harper, K. (2000). *Give me my father's body*. Profile Books.

Jakobsen, U., & Larsen, H. (2024). The development of Greenland's self-government and independence in the shadow of the unitary state. *Polar Journal, 14*(1), 9–27. https://doi.org/10.1080/2154896X.2024.2342117

Kingdom of Denmark. (n.d.) *Greenland: The world's largest island*. Online at: Retrieved February 9, 2026, from https://denmark.dk/people-and-culture/greenland

McGwin, K. (2018). Copenhagen to foot bill to clean up US pollution in Greenland. *Arctic Now*, January, *11*, 2028.

Nielsen, K. H., & Nielsen, H. (2017). *Camp century: The untold story of America's Secret Base under the Greenland Ice*. Columbia Unversity Press.

Petersen, N. (2007). The iceman that never came: 'Project Iceworm,' the search for a NATO deterrent, and Denmark, 1960–1962. *Scandinavian Journal of History, 33*(1), 75–98. https://doi.org/10.1080/03468750701449554

Peterson, O. A. (1951). Eastwind sets new polar record. *Coast Guard Magazine*, January.

Pollard, D. (1956). Men against the Arctic. *Coast Guard Magazine*. February, 1956.

Swain, K. (2024). *Colonel Bernt Balchen: Polar Aviator, Arctic Innovator*. Research and Analysis Division, Ted Stevens Center for Arctic Security Studies.

Team Coco. (2019). *Conan Negotiates with Greenland's Parliament*. Online at: Retrieved February 12, 2026, from https://youtu.be/rE7NtY0daNU?si=BFxbm0paxnQ-4N1O

Tinsley, F. (1950). Rubber Fortresses for A-Bomb Defense. *Mechanix Illustrated, 42*(6), 82–84. 154.

Wager, W. (1962). *Camp century: City under the ice*. Chilton.

Afterword: The American desire for an Arctic island with a more attractive name

In the summer Eirik left to settle in the country he had found, which he called Greenland, as he said people would be attracted there if it had a favourable name.
—*Eirik the Red's Saga*

As the 'Greenlandification' of Greenland proceeded throughout the final decades of the twentieth century and into the 21st, and with it the gradual moves towards self-determination and a reckoning with Denmark, other reminders of Greenland's fraught relationship with the United States also marked the way. These included the mid-1980s visit to Arlington National Cemetery of the Greenlandic descendants of the American explorers Robert Peary and Matthew Henson that dredged up the treatment of Greenland and Greenlandic guides throughout the era of American North Pole ambitions, from Samuel Morton and Elisha Kent Kane through to Peary and Frederick Cook and beyond.

In 2022, a Greenland government commission was established to examine Denmark's historical relationship to the island (Jakobsen and Larsen *op cit.*: 24). This time, the Danish government agreed to join. No similar body has thus far been proposed to make a similar analysis of the historical lust of Americans after a Greenlandic colony. Instead, Greenland remains the place where Americans sought, with precious little reciprocation, fame, riches, museum collections, academic credibility, military security, and natural resources.

Samuel Morton sought to use the Greenlanders as evidence of the racial and cultural superiority of Caucasians and, in the process, earned the title of 'Father of Scientific Racism.' A century after Morton, William Hobbs posited a unified theory of weather over the Greenland Ice Cap that was no more correct but at least possessed the advantage of years of careful field study of Greenland's meteorology. And Hobbs' students would go on to stellar careers themselves as leading American scientific advisors and academic administrators.

Elisha Kent Kane's national significance was recognized in the U.S. by a medal from Congress, a commemorative stamp in 1986 from the U.S. Postal Service, and two U.S. vessels, the destroyer USS *Kane* (DD-235) and the oceanographic research

P. J. Capelotti, *Lost Greenland*, Springer Polar Sciences,
https://doi.org/10.1007/978-3-032-27841-8

vessel USNS *Kane* (T-AGS-27). Internationally, Kane was recognized by the naming of an island in the Arctic archipelago of Franz Josef Land and, 'extranationally,' by the naming of Kane Crater in the high latitudes of the Moon. In Greenland, Kane, if remembered at all, it would most properly be for the graves he disturbed and the grave goods he removed to museum collections in the U.S. That his own grave in Philadelphia would itself one day be vandalized would no doubt strike the Greenlanders as the height of appropriate irony.

Robert Peary has with justice received the most scrutiny of any American polar explorer. For a century, this focus was only on his claim to priority at the North Pole. With the rediscovery of the story of Minik in 2000, Peary's inhumane treatment of the Greenlandic Inuit began to receive long overdue attention, even if Frederick Cook had shouted, during Peary's lifetime, of Peary's abuse of native Greenlanders to anyone who would listen.

Frederick Cook himself experienced the full range of what Greenland and Denmark could do for an American. In less than 6 months in the fall and winter of 1909, he first stashed his polar secrets on Annoatok Island, then raced for Denmark to a rapturous and insanely prosperous reception in Copenhagen, only to suffer the devastating analysis of the University of Copenhagen that his was an obvious fraud. Thereafter he could and would always fall back on a plaintive rejoinder that if he could just return to Annoatok all of his just glory—and what he saw as Pearys's treachery—would be confirmed.

Other Americans, notably Parker 'Shorty' Cramer, staked their careers in daring flights across the Greenland Ice Cap. When he finally succeeded, Cramer was then lost at sea just as he was about to reach Copenhagen and the ecstatic reception that Cook had received but Cramer would have actually earned. And Louise Boyd, whatever praise she earned for her daring expeditions, she deserved far more. Delayed for more than a decade by the secrecy required by the Second World War, her scientific publications set a high standard that no privately funded expedition achieved before or since.

American servicemen by the thousands experienced Greenland during the Second World War and some, like Ted Novak, came away with their world view permanently altered to one of generous and thoughtful cultural relativism. They became the best of the Greatest Generation, returning from their wartime experiences to build a prosperous American middle class, a world-class public education system, a transportation infrastructure that became a model for the world, face down the threat from Soviet communism, and not least rebuild a shattered Europe and set it on course for its own prosperous liberal democracy.

Some achieved their own just global fame. The wartime rescues on the Greenland ice by the magnificent Norwegian-American aviator Bernt Balchen only burnished his already brilliant reputation as the greatest pilot in history over snow and ice. And Balchen then led the massive military operations required to create Greenland as the shield against an expansionist Soviet Russia. With his ever-present slide rule, Balchen was also the first to calculate that Richard Byrd's claim to be first to fly to the North Pole was entirely fraudulent. That Byrd received the Congressional Medal of Honor for a flight Balchen was convinced never took place in the manner Byrd claimed only added to one of the great travesties of polar history. That they ended up buried next to each other at Arlington National Cemetery, just as Matthew Henson's body was removed from its original grave in Brooklyn and placed next to Robert Peary's at Arlington, is yet another irony sprung from the intersections of the reputations they constructed in Greenland (Fig. 1).

Greenland became the front line of the Cold War, with all of the Walt Disney meets Dr. Strangelove manifestations that followed, from secret bases under the ice to proposals for rubber radomes hidden behind inflatable rocks. That history led, on the one hand, to invaluable insights into the history of the global climate via ice core research unimageable even a decade earlier and, on the other, to a legacy of abandoned bases and the military-industrial pollution they left behind. That these abandoned American bases leached toxins and radioactive waste into the Greenlandic environment for decades without the slightest American interest in remediation, is yet another consideration whenever an American public figure tosses out cavalier suggestions of Greenlandic statehood within the United States (Fig. 2).

The archaeological study of these bases has to date found no place within American historical archaeology. Instead, the work has been taken up by European archaeologists, as in the recent study by a team from the Czech Technical University in Prague (Boucek et al., 2023). Such work follows naturally upon the general

Fig. 1 The graves of Bernt Balchen and Richard Byrd at Arlington National Cemetery in the fall of 2025. Byrd's marker announces the Medal of Honor from the flight he claimed to the North Pole in 1927 (photo by the author)

Fig. 2 The author at the site of the abandoned Bluie East 2 base, north of Kulusuk, July 2025

eastern European experience with the refuse left behind by half a century of Soviet defense installations (see, for example, Kiarszys, 2026). It was left to the Danes, under pressure from the Greenlandic government, to finally establish a monitoring station above the environmental time bomb that is the buried ruin of Camp Century, set to emerge into Baffin Bay in mere decades.

The current American interest in Greenland, seemingly based on a suggestion to President Donald J. Trump by an old acquaintance with business interests on the island, seemed shocking at first. But it follows a well-worn American path that would have been all too familiar to American officers like John G. Bourke who witnessed firsthand the displacement of native Americans from their land and the subsequent exploitation of those lands for private profit. Throughout the Plains Indian Wars, Bourke served side by side as an aide to General George Crook as they fought native Americans from the Canadian border to Mexico. Crook's own grave at Arlington, with its stunning bas-relief of the surrender to Crook of the Apaches under Geronimo in the Sierra Madre in 1883, shows just how far a reputation as an 'Indian fighter' could earn a man in eternity. Like the gigantic Peary obelisk in northwestern Greenland, the corporate interests that followed (and often funded) the explorers and soldiers and airmen made sure such men were remembered, and thereby who owed what to whom (Fig. 3).

Trump prefers not to await eternity. As early as 2019 he tweeted an image of a gold Trump Tower looming over a Greenlandic village. When he returned to power in 2025, Trump's Greenland ambition had if anything intensified, egged on by Ronald Lauder. "Beneath its ice and rock lies a treasure trove of rare-earth elements essential for AI, advanced weaponry and modern technology. As ice recedes, new maritime routes are emerging, reshaping global trade and security" (Lauder, 2025). Lauder was quick to explain that his was a beneficent vision of eventual Greenlandic independence, but it was perhaps difficult in this vision to separate that independence from a new American land grab and, with it, a new dependence, built on an American lust for Greenland's apparent trove of treasure.

Fig. 3 Monument to General George Crook, Arlington National Cemetery, fall 2025 (photo by the author)

In all these machinations, little thought has been given to the historical relationship of American commercial ambition to indigenous Greenlandic identity. In a warming world, climate alterations threaten four thousand years of Greenlandic adaptations to the ice and snow. Traditional hunting is under threat from receding ice even as fish and shrimp populations move northwards with the warming of the oceans. A newly industrialized Greenland might generate some local wealth, but almost certainly, if history is any guide, into a very few well-connected pockets, while indigenous identities are trammeled and indigenous peoples left to deal with the environmental fall-out.

As Nicolas Niarchos writes in *The Elements of Power*: "When Trump took power again in 2025, the new government said it was focusing on more concrete dealmaking… In policies that hearkened back to a former era of U.S. policy, Trump wanted the extraction of critical minerals, not just their means of export. He was agitating for them in Ukraine and Greenland as well as Congo" (Niarchos, 2026: 300). In Greenland, however, such mineral extraction had to follow years of exploration on an industrial scale and, as one example, most of the exploratory oil survey holes had thus far come up dry. Not only was the treasure trove still shut tight, it might prove to be largely empty once finally opened.

There is no doubt, however, of the growing importance of Greenland as the planet begins what promises to be a long and extremely fraught adaptation to global climate change. This inexorable human-induced environmental modification is combined with an equally fraught international security situation, where old alliances fray and new challenges in the Arctic from an emergent China and a belligerent Russia both in search of new shipping routes and valuable minerals to transport along them. Even without the prodding of Donald Trump, these factors, along with the continued American investment in scientific research on the island, has once again brought Greenland front and center in the American mind. If what follows is anything like the appreciation of Greenlandic culture discovered by the young U.S. Coast Guard sailor Ted Novak in 1942, then the American mind would be the better for it.

References

Boucek, T., Stará, L., Pavelka, K., & Pavelka, K. J., Jr. (2023). Monitoring of the rehabilitation of the Historic World War II US Air Force Base in Greenland. *Remote Sensing, 2023*(15), 4323. https://doi.org/10.3390/rs15174323

Kiarszys, G. (2026). Past landscapes of bias: Refuse at abandoned cold war Soviet nuclear bases in Poland. *European Journal of Archaeology, 29*(1), 24–44. https://doi.org/10.1017/eaa.2025.10019

Lauder, R. (2025). I'm a Greenland expert—These 3 paths can make it America's next frontier. *New York Post*. February 4, 2025. Online at: Retrieved February 12, 2026, from https://nypost.com/2025/02/04/opinion/im-a-greenland-investor-these-3-paths-can-make-it-americas-next-frontier/

Niarchos, N. (2026). *The elements of power*. Penguin Press.

The manufacturer's authorised representative in the EU is Springer Nature Customer Service Centre GmbH, Europaplatz 3, 69115 Heidelberg, Germany. If you have any concerns regarding our products, please contact ProductSafety@springernature.com

Printed and bound by CPI Group (UK) Ltd, Croydon, CR0 4YY
17/09/2026
02218456-0001